Praise for *Readers Read. Writers Write. Mathers Math!*

For far too long math has been thought of, and taught as, a noun—the facts to be memorized and the procedures to follow. But math is also a verb—it is reasoning, experiencing, wondering. Whereas it is difficult to inspire curiosity and creativity in math as a noun, it is easy in math as a verb. In this book, Deborah Peart Crayton shows us how to make math a verb, how to engage in the "-ings" of math, and how to turn math into something we *do*. In other words, she shows us all how to become mathers.

Peter Liljedahl
Professor, Simon Fraser University
Author, *Building Thinking Classrooms*
Vancouver, BC, Canada

This book is everything! Deborah Peart Crayton names the harm of traditional math practices and offers thoughtful, practical alternatives—with warmth, humor, and heart. We are all mathers—and this book will help you believe it too!

Pamela Seda
Coauthor, *Choosing to See: A Framework for Equity in the Math Classroom*
Atlanta, GA

With straight talk and just enough personality, Peart Crayton suggests ways teachers can bring their expertise in teaching everything else to teaching math. Let's heed her call to help all students know they are mathers who can math!

Pam Harris
Founder, Math is FigureOutAble
Author, *Developing Mathematical Reasoning*
Kyle, TX

This book reminds us that math belongs to everyone. It offers practical ways to shift math instruction so all students feel seen, capable, and confident. Whether you've always loved math or are still figuring it out, this is a thoughtful guide to building math communities where students thrive and math becomes just as natural as reading and writing.

Graham Fletcher
Math Specialist
Atlanta, GA

Deborah Peart Crayton has crafted a book that is as transformative as it is practical. *Readers Read. Writers Write. Mathers Math!* is an essential read for elementary educators, instructional coaches, and school leaders committed to collectively shifting the current narrative that math is optional, beginning with our youngest learners. By integrating literacy-based approaches and real-world applications, Crayton provides a path forward for making math a joyful, natural, and necessary part of learning. This book is not just about improving math instruction—it's about reshaping mathematical identities for generations to come.

Dionne Aminata
Founder and CEO, Math Trust
Emeryville, CA

Many educators feel unsure about math, while others love it and want to share that joy. The Mather Movement speaks to both. Whether you're working to overcome your own math story or hoping to inspire others, Deborah Peart Crayton offers powerful tools for reshaping your relationship with math. With empathy and wisdom, she provides a path forward: one that builds confidence, breaks old cycles, and helps you grow as both a learner and a teacher.

Robert Kaplinsky
President, Grassroots Workshops
Long Beach, CA

Finally, someone's calling it like it is: Math is just as essential, useful, and FUN as reading! If you're ready for a revolutionary approach to teaching that shatters the myth of the "math person" and helps reduce math anxiety, you've found it!

Vanessa "The Math Guru" Vakharia
Author, *Math Therapy: 5 Steps to Help Your Students Overcome Math Trauma and Build a Better Relationship With Math*
Toronto, Ontario, Canada

As a professional learning provider, this book is written in ways that I think could enhance the teaching practices for elementary and middle school math teachers. As a mother to young mathers, I have learned accessible ways to continue to include math in our every day outside of formal games and counting. I can't wait to try body-gons at home with my mathers.

Ayanna Perry
Director, Outreach and Dissemination, Knowles Teacher Initiative
Bowie, MD

A must-have for all educators who want their students to see themselves as mathers! Deborah Peart Crayton provides us with a lovely tool that leverages her voice and experience to help us become better Mathers ourselves while helping our students come to see themselves as Mathers.

Zak Champagne
Chief Content Officer, Flynn Education
Olympia, WA

This book invites educators to teach mathematics with care, creativity, and purpose. Through inspiring interviews with women in math and empowering "mathfirmations," Peart Crayton reminds us that joy, brilliance, and belonging are at the heart of learning. This book nurtures teacher agency and equity-driven practice, and after reading, you'll proudly proclaim, "I'm a mather!"

Desiree Y. Harrison
Elementary Instructional Coach
Farmington, MI

Deborah challenges us to connect our strengths in literacy instruction and consider how to apply that to math instruction. Grounded in real classroom practice, she demonstrates how storytelling, discussion, and language-rich routines can foster positive math identities and help every child see themself as a mather. It is a call to action to change our culture's negative attitude to mathematics. A powerful resource for everyone who strives to make math meaningful and empowering.

Marria Carrington
Director of Math Leadership Programs, Mount Holyoke College
South Hadley, MA

This book is a much-needed invitation to reimagine what it means to teach and learn math. Deborah writes from the heart and shares her insights and practical strategies to build math communities where all learners can thrive. If you are committed to nurturing joyful, inclusive, and empowering math experiences for all students, this book has everything you need.

Mike Flynn
CEO, Flynn Educational Consulting INC
Northampton, MA

READERS *Read.*
WRITERS *Write.*
MATHERS *Math!*

I dedicate this book to my children, Naomi, Nehemiah, and Naja, who shared me with all the students I have ever taught and to the Village who helped me raise them. I also dedicate this labor of love to my hubby, my friend, my Chief Support Officer, Derrick, who has supported me every step of the way.

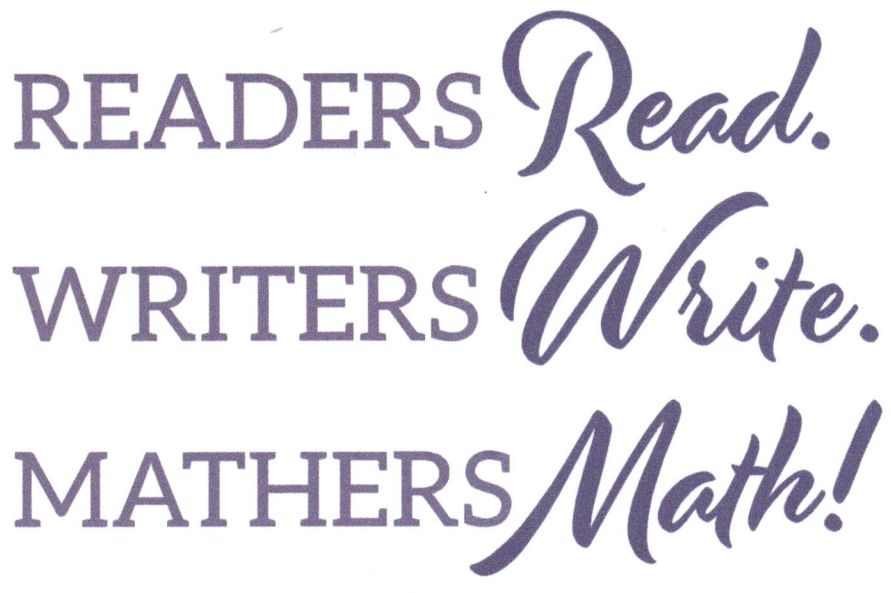

Bridging the Gap Between Literacy and Mathematics

DEBORAH PEART CRAYTON
Foreword by Jennifer M. Bay-Williams

CORWIN

CORWIN

FOR INFORMATION:

Corwin

A SAGE Company

2455 Teller Road

Thousand Oaks, California 91320

(800) 233-9936

www.corwin.com

SAGE Publications Ltd.

1 Oliver's Yard

55 City Road

London EC1Y 1SP

United Kingdom

SAGE Publications India Pvt. Ltd.

Unit No 323-333, Third Floor, F-Block

International Trade Tower Nehru Place

New Delhi 110 019

India

SAGE Publications Asia-Pacific Pte. Ltd.

18 Cross Street #10-10/11/12

China Square Central

Singapore 048423

Vice President and
    Editorial Director: Monica Eckman
Senior Acquisitions
    Editor: Debbie Hardin
Senior Editorial
    Assistant: Nyle De Leon
Production Editors: Nicole Burns-Ascue,
            Veronica Stapleton Hooper
Copy Editor: Heather Kerrigan
Typesetter: C&M Digitals (P) Ltd.
Proofreader: Lori Newhouse
Graphic Designer: Gail Buschman
Marketing Manager: Margaret O'Connor

Copyright © 2026 by Corwin Press, Inc.

All rights reserved. Except as permitted by U.S. copyright law, no part of this work may be reproduced or distributed in any form or by any means, or stored in a database or retrieval system, without permission in writing from the publisher.

When forms and sample documents appearing in this work are intended for reproduction, they will be marked as such. Reproduction of their use is authorized for educational use by educators, local school sites, and/or noncommercial or nonprofit entities that have purchased the book.

All third-party trademarks referenced or depicted herein are included solely for the purpose of illustration and are the property of their respective owners. Reference to these trademarks in no way indicates any relationship with, or endorsement by, the trademark owner.

No AI training. Without in any way limiting the author's and publisher's exclusive rights under copyright, any use of this publication to "train" generative artificial intelligence (AI)or for other AI uses is expressly prohibited. The publisher reserves all rights to license uses of this publication forgenerative AI training or other AI uses.

ISBN: 978-1-0719-4913-9

Library of Congress Control Number: 2025020323

DISCLAIMER: This book may direct you to access third-party content via web links, QR codes, or other scannable technologies, which are provided for your reference by the author(s). Corwin makes no guarantee that such third-party content will be available for your use and encourages you to review the terms and conditions of such third-party content. Corwin takes no responsibility and assumes no liability for your use of any third-party content, nor does Corwin approve, sponsor, endorse, verify, or certify such third-party content.

CONTENTS

Foreword — xiii
Preface: Mathers 4 Life — xvii
Acknowledgments — xxiii
About the Author — xxv
Featured Mathers — xxvii

INTRODUCTION: IS MATH OPTIONAL? — 1
   The Legacy of Gatekeeping — 2
   Who Is This Book For? — 3
   How Can This Book Help? — 3
   How Is This Book Organized? — 4
   How Can You Use This Book? — 5

PART 1 • READERS READ. WRITERS WRITE. MATHERS MATH! — 7

1 WE ARE ALL READERS, WRITERS, AND MATHERS! — 8
Reading and Writing for All! Math for the Few? — 9
   Approaching the Teaching of Math Like the Teaching of Reading and Writing — 10
   How Did We Get Here? — 12
Introducing Math-ers — 13
Let's Redefine the Core Academic Skills! — 15
Learning Math for Life — 20
Where's the Math in That? Musicians Are Mathers! (Ashley Cuthbertson) — 22
Time to Reflect and Take Action — 28
Mathfirmation — 29

2 REDEFINING THE CORE ACADEMIC SKILLS — 30

Literacy Leads the Way — 31
Early Literacy + Early Numeracy = Academic Success — 32
Counting Is as Easy as 1-2-3 . . . Or Is It? — 34
    1. Standard Order Principle: Knowing the number sequence and understanding that it matters — 36
    2. One-to-One Correspondence Principle: Paring one number name to one object as you count — 36
    3. Cardinality Principle: Understanding the number of elements that are in a set and understanding that the last number said when counting is the value of the set — 37
    4. Conservation of Cardinality Principle: Understanding that even when the order or arrangement of the set of objects changes, if none were added or taken away the total remains the same — 37
    5. The Successor Principle: Understanding that there is a number that follows each number that is 1 more — 38
Get Counting! — 41
The Habits of "Good" Learners — 44
Where's the Math in That? Bakers Are Mathers! (Sara Fludd) — 48
Time to Reflect and Take Action — 53
Mathfirmation — 53

3 LET'S SHARE STORIES NOT PROBLEMS! — 55

What Makes a Story a Story? — 56
Oral Traditions — 58
Visual Storytelling — 60
    Using Contexts to Support Conceptual Understanding — 62
    Visualizing for Mathematical Understanding — 66
Where's the Math in That? Yogis Are Mathers! (My Story) — 79
Time to Reflect and Take Action — 83
Mathfirmation — 83

PART 2 • MATHERS GONNA MATH!: TAKING MATH OUTSIDE OF THE MATH BLOCK — 85

4 MATHEMATIZING ACROSS CONTENT AREAS — 86
Mathing Outside of the Block — 87
Highlighting the M in STEM — 89
Science and Math Go Together — 92
The Story the Data Tell Us — 93
- Would You Rather . . . ? — 95
- Colorful Candies or Dangerous Moon Rocks? — 97

Slow Reveal Graphs: Watching the Story Unfold — 100
Mathing Is a Social Endeavor — 102
Mathers Advocate for Fairness — 104
Where's the Math in That? Entrepreneurs
   Are Mathers! (Brittany Rhodes) — 106
Time to Reflect and Take Action — 113
Mathfirmation — 113

5 DON'T SKIP THE FUN STUFF! — 115
Fingers Are Made for Counting! — 116
Which Way Is Up? — 120
Polygons or Body-Gons? — 122
Do You Know What Time It Is? — 126
Show Me the Money! — 129
Focus on the Fun Stuff — 134
Where's the Math in That? Triple Threats
   Are Mathers! (Rose Jackson Moye) — 134
Time to Reflect and Take Action — 138
Mathfirmation — 139

PART 3 • BUILDING A COMMUNITY OF READERS, WRITERS, AND MATHERS — 141

6 BECOMING THE MATHER YOUR STUDENTS NEED YOU TO BE — 142
Be the Anchor! — 144
Comfort B4 Confidence: A Framework for Change — 146
Building Comfort: Let's Do the Math! — 148
Building Competence: Understanding the Math We Teach — 152
Building Confidence: Thinking Flexibly and Affecting Change — 154

Building Community Through Collaboration: Shifting the Culture	157
Math Without Borders: We Are All Mathers!	160
Teachers Must Lead the Way!	162
Where's the Math in That? Coaches and Trainers Are Mathers! (Angelina Perrone)	163
Time to Reflect and Take Action	169
Mathfirmation	170

7 CREATING THE CONDITIONS FOR POSITIVE MATH (LEARNING) EXPERIENCES — 171

It Takes a Village . . .	172
We Belong Together!	176
Welcome to the Math Circle!	180
Mathing as a Sensory Experience	182
We Read. We Write. We Math. Together!	185
Where's the Math in That? Artists Are Mathers! (Naja Brooks)	186
Time to Reflect and Take Action	190
Mathfirmation	191

8 SUMMING IT UP! — 193

Where Do We Go From Here?	194
Wouldn't You Like to Be a Mather Too?	195
Be the Change!	196
Hope for the Future!	198

References	201
Index	205

FOREWORD

Gates and Gatekeeping

By Jennifer M. Bay-Williams

Whether you already are a big fan of the Mather Movement or you are just reading this phrase for the first time, this book will inspire you to be better at inviting everyone into mathematics. *Readers Read. Writers Write. Mathers Math!* is a powerful call for a cultural shift in mathematics in terms of how it is *perceived* and *received*. Through poetic writing, contrasts to reading and writing, and relatable examples, Peart Crayton illuminates what has led to mathematics being a gatekeeper and how we (the readers) can open that gate. In this foreword, I will briefly highlight ideas from the book related to these two constructs (and why they matter).

Gatekeeping comes in the form of (mis)perceptions about mathematics. Perceptions about mathematics, as Peart Crayton beautifully describes, are quite different than those around reading and writing. For example, a common perception is that reading and writing are necessary for everyone, regardless of career interests, while mathematics is considered a necessity for only *some* people in only *some* careers. Peart Crayton attends to numerous (mis)perceptions throughout the book, including that math is "not creative" and "some people are born with a math gene." Collectively, these (mis)perceptions have led to a culture wherein it is okay to say, "*I am not good at math*." Students who do not think they are good at math may not persevere on complex tasks, or pursue advanced math courses or math-related careers. These (mis)perceptions are gates that deny access to mathematical proficiency. It is through shifts in how we think about, talk about, and engage in doing mathematics that we can open these gates. For example, offering "mathfirmations" to students can change the way they feel about mathematics. As Peart Crayton writes, "Let's invite them with our words and our actions to be *mathers*."

How mathematics is *received* has resulted in the perceptions highlighted on the previous page (and many others discussed in this book). Teachers are gatekeepers, and thus, it is our job to figure out how to open the gates for all of our students. Mathematics is often received (taught) by watching the teacher do the math and then remembering what was shown. Mathematics is received in isolation from other topics. Mathematical tasks, even when they have a context, are not connected to students' interests or experiences. These teaching actions lead students to say, *"I don't like math,"* and thus keep the gates closed for too many students. When mathematics learning is meaningful, connected, and engaging, students develop confidence and competence in mathematics—they feel like a mather! This book is packed with content tips and rich tasks that offer such gate-opening ideas. For example, using oral storytelling instead of classic word problems, connecting to other disciplines, and doing math outside of the math block.

The messages in this book attend to all mathematics topics, from counting to geometric shapes. And, a strong connection for me relates to developing fluency. Certainly our cultural obsession with learning algorithms has contributed to the perceptions people have about mathematics, but that does not have to be the case. Creativity and decision-making are at the core of mathematical fluency. For example, as children think about how they might solve 32–19, they might make sense of the problem by drawing a picture, building something with manipulatives, or putting it in a context (create a story). Students might consider how they want to think about the problem, for example, by asking, do I want to count back (take away) or find the difference (compare) to solve the problem? Once this decision is made, the next decision is about how to jump back or up. As students engage in this reasoning, they develop competence and confidence in various ways to subtract. And, developing fluency can be fun. Peart Crayton notes that "There is a pervasive belief that mathematics by definition is not supposed to be fun that starts in elementary school and continues well into the upper grades and adulthood." This resonates with me, as so much of fluency instruction is rote practice. Yet, games provide substantial and enjoyable practice wherein students are able to think aloud to share their reasoning and to listen to and learn from their peers. Such play is joyful and builds their competence. Thus they engage as a mather and this supports their journey to procedural fluency. As Peart Crayton suggests, "Find time . . . to focus on math through games and other fun activities."

Peart Crayton brings her life experiences and her joy of life, reading, writing, mathing, and learning to this book. This includes many stories and ideas from her 30+ years of teaching and leading. She is *the* Queen Mather and is on a mission to ensure that everyone recognizes that they are also mathers. Her background in literacy, early childhood, and communication are apparent in the beautiful way she brings out the human experience in the learning of mathematics.

Readers Read. Writers Write. Mathers Math! is an inspirational book, but it offers us more than inspiration. It is a call for much-needed change that helps us reflect on what math learning has looked like (and the consequences of this) and what it can look like (and the benefits of this). Peart Crayton notes, "There is nothing like witnessing Pure Math Joy!" True. And, this book provides you with a wealth of resources to ensure that you get to witness this joy in your own mathing and with your own students.

PREFACE

Little Debbie:

People often ask me, if I am a mathematician.

My immediate response is, "Well that's just simply fiction."

Mathematician is a very fancy label

But I myself was never invited to that table

Even though math came easy to me

Each year teachers chipped away at my curiosity

Don't ask questions, just follow the steps with ease

Make the grades without understanding, simply perform and appease

Video of Deborah performing a version of this preface.

https://qrs.ly/kjgnms8

Math is only for those with the imaginary math gene and gender is a factor

Let's be honest, sometimes to be the star in math class, you just need to be the best actor

And then we wonder why people say with pride

Who me? I am NOT a math person, all of my math dreams have shriveled up and died

Mathematicians study math to solve the problems of the world

But anyone can choose the path and anyone can put in the work

Yes anyone, yes even a girl

However, there must be a distinction between math for a career and math for life

Because math is all around us and shouldn't cause us any strife

I mean, scientists, dancers, construction workers, all mathing, am I right?

A reader reads and a writer writes, there is no opting out

We read and we write for grades, but that's not what it's all about

Readers and writers get to use their imaginations

And being literate in America, determines our life stations

The truth is math is not just a subject but an experience to be had

And maybe if there were more joy in math class, then math class wouldn't be so bad

And if we helped students connect the math to their lives, then they wouldn't be so darn sad

Because we are all born with math intuition and mathematical ideas; and that's a fact

Just ask any toddler to choose between 2 or 5 of their favorite snack

Our brains are wired to recognize patterns and solve problems in all different ways

READERS READ. WRITERS WRITE. MATHERS MATH!

We face puzzles and logical challenges and reason our way through them on most days

We code, we sketch, we plan, and design the towers of our dreams

Yet math is never viewed as the solution, only the problem in all of the memes

We cannot fix it all today, this problem is way bigger than you and me

But, there is something we can do to get math where it needs to be

And it starts with disrupting elitist gatekeeping practices that are our legacy

So here we are, with the biggest math problem left unsolved

How do we reclaim mathematics and get everyone involved?

Can we balance the scales and prove that as humans we have truly evolved?

First, let's change our language and stop all of this Us versus Them

We don't all have to choose a career in mathematics or in the field of STEM

And educators, this all begins with you, you must believe that you are mathers too

What's a mather?

Well, a mather is a person using math to make sense of the world you see

That's you, your mama, your students, your friends,

And of course, the Queen Mather, well, that's me

Math deserves a verb, a makeover, and a brand-new chance

To be seen as inclusive and necessary even when we dance

Can you commit? Will you decide to advocate for Mather Pride?

And be the one to save math dreams that came to school and died?

Paint the picture, help others see Math's softer side and creativity?

Math for all, math for life, math for justice and liberty!

Because we say math is liberation, but liberation requires us to resist

We must stand against inequitable policies and practices

Instead of just telling kids they must persist

Of course, productive struggle leads to learning that will last

Key word, productive NOT just struggle

Can we leave the bootstraps in the past?

I implore you one and all to join the Mather Movement with me

Stand up, stand out, and be the mathers our students need us to be.

So, what's it gonna be? What will you choose to do?

I know I'm a mather! Wouldn't you like to be a mather too?!

 My Mathematical Mind Team

From left to right, Adrienne Baytops Paul, Tisha Jones, and Deborah Peart Crayton, the Queen Mather

Source: Deborah Peart Crayton

Writing a book with practical strategies for educators made me think about the many conversations I've had with colleagues in the break room or hallways and in the cozy corners tucked away in conference centers between sessions. However, I am also a current doctoral candidate learning the nuances of academic writing. That said, I worked to strike a balance between conversational and informational tones in this book. At some points in the book, you will notice citations, research, and more formal writing, but in many places, you could likely imagine us having a friendly conversation about teaching and learning over coffee. I embrace both as a part of my academic and professional identity and hope that the message is not lost in translation.

I consider myself a word nerd and a lover of language, so my goal as a writer is to convey meaning, tell stories, and paint beautiful pictures with words to evoke emotional responses and provide vivid imagery for the reader. I also recognize that words have power and can be interpreted differently by each individual. My word choices were guided by my journey, my background and experiences, and the settings in which I have taught in the past 30+ years. I acknowledge that my choice of words might be interpreted in ways I may not have intended, and take ownership of the impossibility of perfection.

ACKNOWLEDGMENTS

There are many characters in the life and story of Deborah Peart Crayton, so it would be impossible to thank everyone by name who has ever planted a seed, shared an encouraging word, listened to me go on and on about my passions, or leaned in as I read chapters aloud. This book is a culmination of all of the experiences I have had inside and outside of the classroom, so to any teacher, parent, student, or friend who sees yourself in this story, I say thank you for the inspiration. This book would not have been possible without my students and their families who trusted me to try new approaches, celebrated our wins, and moved past our losses, so a huge thank you to all of them.

Family means the world to me, so I must acknowledge my parents Derrick and Valerie Peart for the sacrifices they've made so I could spread my wings and fly. I am grateful for the lessons my siblings taught me about perseverance and the stories we've written together. Thanks to my Crayton family who welcomed me with open arms and encouraged me to focus on the prize. I thank my children who have always been my greatest cheerleaders and my favorite students. And, of course, I acknowledge my wonderful husband, Derrick, and our fur babies, Dolly, Daisy, and Jax, for understanding when I had to write for hours. I am indebted to them for their love and patience.

I acknowledge My Sister Circle for the love and support they have shared over the years and extend gratitude to my fellow members of the Mather Movement who have cheered me on, spoken my name in rooms where I was not present, and shared opportunities with me. Special thanks to Niky and Layla's Curve for our Mather Merch that has made the **#MatherMovement** visible.

Thanks to Debbie Hardin, my amazing editor, who took a chance on a girl from the Bronx with a story to tell, and to Nyle who is always working behind the scenes to move things along. I am honored and humbled by your unwavering faith in me to get the job done.

This book is inspired by you all and the greatest gift I have ever given.

Corwin gratefully acknowledges the contributions of the following reviewers:

Dionne Aminata
Founder and CEO, Math Trust
Emeryville, CA

Zak Champagne
Chief Content Officer, Flynn Education
Olympia, WA

Susie Katt
K-2 Mathematics Coordinator, Lincoln Public Schools
Lincoln, NE

Christina Lincoln-Moore
Elementary Mathematics Coordinator, Los Angeles County
    Office of Education
Inglewood, CA

Rosamaria Murillo
Principal, La Habra City School District
Whittier, CA

Ayanna Perry
Director, Outreach and Dissemination, Knowles Teacher Initiative
Bowie, MD

Venessa Powell
Education Consultant, TeachEDX
Manchester, Jamaica

Crystal Watson
Principal, Cincinnati Public Schools
Cincinnati, OH

ABOUT THE AUTHOR

Deborah Peart Crayton is the founder and Queen Mather of My Mathematical Mind. She started the **#MatherMovement** to disrupt the idea that math is optional. Deborah is a sought-after keynote speaker and dynamic elementary education consultant. With over 30 years in the field of education, Deborah speaks on a variety of topics related to math identity, elementary math content and instructional practices, and literacy connections to mathematics. She has shared her message at the local, regional, and national levels at state, district, and community events, and educational conferences.

My Mathematical Mind
https://qrs.ly/h7gnms9

Deborah is an expert at creating inviting and nurturing environments to cultivate positive mathematical identities, and has served in that capacity as a mentor and coach for elementary educators for over 20 years. She has had success with adapting instruction while maintaining grade-level expectations and possesses a deep knowledge of instructional practices and frameworks to engage all learners, including Cognitively Guided Instruction, Math Recovery, the Orton-Gillingham Approach, and Mathematical Language Routines.

Deborah holds a bachelor's degree in speech communications and early childhood education, a master's degree in educational studies with a concentration in literacy, graduate endorsements in elementary mathematics and teacher development, and is pursuing a doctorate degree in education program development and innovation. Her research interests are focused on the connections between math anxiety and math teaching efficacy and effective practices for developing high-quality professional learning. Deborah has dedicated her career and doctoral studies to supporting educators with innovative teaching strategies that allow students to see themselves as assets to the learning community and curious problem solvers. Deborah believes that all children deserve high-quality instruction and the opportunity to become competent readers, writers, and *mathers*.

FEATURED MATHERS

Musicians Are Mathers: Ashley Cuthbertson

Source: Ceylon Mitchell

Ashley Cuthbertson, EdM, NBCT (she/her) is a nationally recognized arts consultant, speaker, and author of *Music As a Vehicle: A Practical Guide to Implementing Culturally Responsive Teaching in Today's Music Classrooms.* Her work explores how the arts are a vehicle for building the skills needed for success in school, career, and life. Previously, Ashley was a dedicated music educator, teacher leader, and adjunct professor with extensive experience in diverse school settings. Today, she partners with schools and districts, supporting them in reimagining their arts programs through a culturally relevant lens that builds critical skills for success both inside and outside the classroom. She also works with organizations to help them integrate arts-based strategies to strengthen team culture and leadership.

Learn more about Ashley Cuthbertson

www.Ashley Cuthbertson .com

Bakers Are Mathers: Sara Fludd

Source: Boyzell Hosey

Sara Fludd is the founder and Queen Waffler of Pop Goes the Waffle. She is known across Florida for her sweet and savory treats. She is a waffler, a baker, and a mather.

Entrepreneurs Are Mathers: Brittany Rhodes

Learn more about Brittany Rhodes

https://mathequalsme.com/

Brittany Rhodes is the founder and General MATHager of Black Girl MATHgic and Math = Me, where she curates supplemental K-8 math confidence kits, workshops, and professional development to reduce math anxiety and strengthen math confidence in students and educators. With a bachelor of science degree in mathematics from Spelman College and an MBA from Carnegie Mellon University, Brittany's work has been featured in *Black Enterprise, Good Morning America,* and *Forbes.* A proud Detroit native, she is dedicated to closing the math confidence gap and empowering the next generation of math enthusiasts.

Triple Threats Are Mathers: Rose Jackson Moye

Rose Jackson Moye is a gifted actress/dancer/singer, who has graced stages from London to Los Angeles. Along with several stints in regional theater and on Broadway, Rose has had several roles in both television and film. Rose has also been an energetic and engaging dance instructor for over 26 years. Rose has never met an audience she did not love. However, Rose's favorite audience consists of her husband, retired TV Producer Michael G. Moye, and her favorite psychologist, daughter Memphis.

Source: Bob Capazzo

Coaches and Trainers Are Mathers: Angelina Perrone

Angelina is native New Yorker who ventured to the Carolinas to play both field hockey and lacrosse at Belmont Abbey College. She graduated with a business degree and double minored in sports management and criminal justice. Later, Angelina attended Wingate University where she was the graduate assistant for the women's lacrosse team while pursuing a master of business administration degree. She now serves as a collegiate lacrosse coach at Catawba College and a trainer at Burn Boot Camp Belmont. During her free time, she loves to travel, work out with friends, read books, and enjoy her favorite artists live in concert.

Artists Are Mathers: Naja Brooks

Naja Brooks is a freelance illustrator with a wide range of work and an impressive multimedia portfolio. She completed her undergraduate studies at Savannah College of Art and Design and is currently pursuing graduate studies in digital media at Kennesaw State University. Even though math is not something immediately thought of when it comes to the creative arts, it is something Naja has always enjoyed. Naja is the creator of the Mather Kids and the illustrator for this book.

Learn more about Naja Brooks

https://WorkByNaja.myportfolio.com

INTRODUCTION
Is Math Optional?

When was the last time you went out with friends and someone asked you to read the menu because they said they weren't reading people? I'll bet your response is never. If I ask a similar question with mathematics at the center, I feel confident the answer would be quite different. Intellectuals toss their credit cards into the center of the table expecting that the one "math person" in the group will do all the calculations and let everyone know what they owe. Why is it acceptable for someone to publicly admit that they are not a "math person" when it is time to split the check or calculate the tip for a large group but this same person would never publicly proclaim that they needed someone else to read the menu to them? It might have something to do with the pervasive idea that some people are born with math brains while others are not.

This is unequivocally false.

We are all wired to do mathematics, so why do so many people reject mathematics as a part of their identity? Why do so many citizens view mathematics as something they don't need to function in society? More importantly, what can we do to solve this math problem? First, we need to acknowledge that math is NOT the problem.

Many people can remember the moment they fell in love with mathematics, but even more individuals have a story ready to share about why they fell out of love with mathematics. My earliest memories of engaging with mathematics happened long before the first day of my formal education. I was curious about how things work and often designed patterns using blocks, drawings, or food items. Solving problems and asking questions was a daily occurrence for me. School was the place where I thought I might find the answers to my questions, but by fourth grade I realized that asking too many questions was sometimes perceived (even by a teacher!) as being sassy, rude, or disrespectful. In my quest to be a model student, I yielded to the pressure and assimilated, stopped asking questions, and instead learned math passively, following steps and prescribed procedures to achieve success. Unfortunately, many have similar stories of their earliest classroom math experiences that shaped their math identities as adults.

Historically, public education focused on literacy skills for all and mathematics for some. It was believed that common folks, the laboring class, had no need for advanced mathematics or deep conceptual knowledge, so math instruction was limited to basic computation skills. This is why the core academic skills known as the Three Rs—reading, writing, and 'rithmetic—became the focus in the common schools. Math was taught using the rule method: A rule for a problem was introduced, the rule was memorized, and then came drill and practice. Only privileged boys aged 12 and older were given access to high-level mathematics in preparation for the path to college. Some believed that girls and women were too "delicate" to study mathematics since it was such a challenging subject. This legacy of discriminatory practices in mathematics lives on, and the gatekeepers stay poised to redirect anyone who doesn't fit the stereotype yet is bold enough to believe they can achieve success in the field of mathematics or other STEM careers to exit stage left.

Unfortunately, the expectations for success with mathematics continue to be high for a select group and virtually nonexistent for others. In schools, there is still an emphasis on literacy, and teachers are supported more heavily through professional development to improve literacy instruction. In society there is a goal for everyone to be literate, and adult literacy programs are offered to improve literacy skills. Over time, the criteria for being literate has changed, but in general being literate in the United States is having the ability to use written or printed information to function in society, develop knowledge, and achieve one's goals. Most jobs often require a high literacy level, and the topic of adult literacy is researched extensively. This level of support for improving math skills and expanding math content knowledge just isn't there.

A 2019 report by the National Center for Education Statistics determined that mid to high literacy in the United States was 79 percent with 21 percent of US adults categorized as having "low level English literacy." But I wonder what our numbers would look like if to be considered literate in the United States, you also had to be numerate, a problem solver, and good at mathematical reasoning. What if we redefined what it means to be literate as the ability to read, write, speak, and listen in a way that lets us communicate effectively **and** to use mathematics to make sense of the world?

This book is written for the teacher who is nervous about teaching mathematics. If you identify as someone who is math anxious but would like to explore options for elevating your math teaching, you will find strategies to support you on your math journey as you develop your math identity and become the mather your students need you to be. It is also for the teacher who is confident teaching mathematics. If you recognize that you have always loved math so you wonder why others can't see its beauty, it is important to consider the experiences that supported your math identity development to create similar positive math experiences for students. In this book, you'll find strategies you can use to create a sense of belonging and engagement to help your students thrive.

This book is also written for coaches and instructional leaders who have a math vision they hope their teachers will be prepared to strive toward. If you want to nurture positive math identities in students, it begins with your faculty and staff. You will find strategies for developing a math culture in which adults model a growth mindset (Dweck, 2016) and share beliefs that cultivate math joy with their students and the community beyond the classroom. Honestly, this book is for any educator who is invested in giving all students access to high-quality math instruction and a lifelong love of math learning.

This text provides strategies for reclaiming mathematics, as adults, to address the societal issue of separating literacy skills, which are needed for life, from math learning, which is presented as needed for

STEM careers. By redefining the core academic skills and rebranding mathematics as necessary for everyone—even those who do not go on to STEM careers—reading, writing, and mathing become standard. Mathematics is a necessary part of our lived experiences and can be enjoyed if we understand its multifaceted nature.

This book is divided into three parts. Part 1, *Readers Read. Writers Write. Mathers Math!*, shares evidence from neuroscience that dispels the myth of the "math brain" and explores practical ways in which we can capitalize on the natural math intuition that we are all born with. In this section, a brief history of the origins of elitism in mathematics is highlighted to expose the legacy of discriminatory practices in math spaces. By proposing alternative instructional practices and sharing practical tips, we establish the groundwork for changing the narrative about what it means to be good at math. We examine the habits of good learners and celebrate the power of storytelling across disciplines as ways to support students and adults with overcoming the angst that exists when facing word problems. The section ends by leveraging structures and practices used for literacy instruction, to empower teachers to lean into their strengths and bring joyful experiences into their math instruction. By mathematizing literature, students can learn more about using mathematics to make sense of the world and connect math to their lives. One day at a time, we can strive to change our language and redefine the core academic skills to embrace reading, writing, and mathing as necessary skills for success in school and life.

Part 2, *Mathers Gonna Math!: Taking Math Outside of the Math Block*, focuses on building bridges between content areas and highlights ways for students to connect mathematics to everyday tasks and experiences. We demonstrate the ways in which math supports science exploration and how essential mathematics is to developing compassion toward one another and affecting change. This part of the book zooms in on ways in which mathematics is present in places and ways that are often overlooked, like reading a map, solving logic problems, or noticing patterns. It is important to help students celebrate the ways they are already mathing every day to disrupt the narrative that math is optional. Don't skip the fun stuff when time is short! We discuss the need for students to see mathematics as more than just a class subject. In this part of the book, mathematics comes to life, as we focus on math in action and how mathematics can be used as a tool for solving not only **word** problems but **world** problems.

Part 3, *Building a Community of Readers, Writers, and Mathers,* shifts focus to the work of creating safe math learning spaces for students. We explore strategies that move math mindset work from the bulletin boards to our classroom culture. As educators create the conditions for positive math experiences, students feel empowered to lead their learning and lean on classmates as they develop collective agency. Children are born with math intuition and have dreams that could be supported if we help them develop their sense-making superpowers. To create thriving math communities, there are necessary shifts that adults need to make to ensure they are representing mathematics in a holistic way, equipped with strategies for alleviating math anxiety in themselves and their students. As we examine our own math stories, rewriting negative narratives we have replayed for years, we explore alternatives for math instruction that have the power to nurture positive math identities in our students.

Part 3 also focuses on teacher experiences and collaboration. To create joyful math experiences for students, teachers must know what it feels like to experience math joy. I introduce the Comfort B4 Confidence Framework as a tool for guiding educators as they hone their craft. As we develop our comfort with mathematics, confidence will follow. This section highlights the benefits of exploring new approaches for teaching mathematics and offers powerful solutions for building a community of mathers. As teachers find safe spaces where they can be vulnerable, adult learning can happen. Adults who engage in math play begin to see math in a new light and can anticipate student strategies together, as they do the math of the lessons they teach. Through professional learning and purposeful planning, educators sit in the seat of learner, teacher, and analyst and sharpen their lenses as they get curious about student thinking. We must see ourselves as mathers, understand the math we teach conceptually, and model a growth mindset for our students. Learning mathematics won't always be easy, but our students need to know that on the other side of productive struggle is victory. We have a responsibility to lead the way!

This book can be used to spark conversations about what it means to be good at math. It can be read in order, one chapter at a time, or it can be used as a quick reference to find tips, activities, historical tidbits, and connections between mathematics and the world. Readers can use this book to lead a book study or guide professional learning to support math identity development and enhance mathematics instruction by leveraging the strengths teachers bring from teaching literacy. This book is a tool for disrupting the current narrative that math is not for everyone and establishing the counter narrative that reading, writing, **and mathing** are necessary to function in society. Starting with educators, this book guides adults through a process for reclaiming mathematics as a part of their identities and building up the confidence to explore mathematics in new ways, so they feel empowered to create thriving math communities for their students.

Through the *Where's the Math in That?* feature at the end of each chapter, we connect math to art, music, dance, and more. Educators are equipped with practical conversation starters and real-life examples for expanding their students' view of what it means to be a mather for life. Through interviews and personal stories about the role math has played in shaping lives, educators hear different perspectives of why math is necessary across disciplines. For the students who don't believe they need math because they plan to be a rock star, professional athlete, or entrepreneur, these testimonials equip teachers with some of the reasons why, no matter the career path we choose, math is a necessary part of the journey. This feature provides the simple answer to the question, "Where's the math in that?" Right here!

You can read this book to learn practical strategies for cultivating curiosity and joy in math classrooms and for creating the conditions for shifting mindsets and beliefs around who is born to do mathematics. Math isn't optional! Together we can rebrand mathematics and build a community of competent readers, writers, and **mathers.**

PART 1

READERS READ. WRITERS WRITE. MATHERS MATH!

WE ARE ALL READERS, WRITERS, AND MATHERS!

Source: istock.com/JohnnyGreig

Have you ever seen a friend cringe at the mention of filing taxes, as they proclaim, "I am not a math person"? Have you ever seen parents avoid reviewing their child's math homework, sighing, "I am not a math person"?

But we don't hear adults say, "I am not a reading person." In fact, adults who have reading challenges typically go to great lengths to hide the fact that they struggle with reading. Whether they fear shame and ridicule or believe that reading is critically important to their success, no one boasts that they cannot read or write. To be literate—that is, able to read and write—is perceived as being educated. The definition of *literate* clearly articulates that an adult needs to be able to use printed or written information to function in society. But what does it say about society's perception of math if many of us are willing to admit—sometimes proudly—that we aren't good at math?

Reading and Writing for All! Math for the Few?

Where does this acceptance of the idea that math is optional come from?

In classrooms, we call students readers and writers, which clearly involves the students in ownership of the doing of the reading and writing. But when it comes to math, it is almost as if math is done *to* students. Classrooms typically don't emphasize the ownership of being a math person by labeling those who do math as "mathers."

There has been much written about who belongs in math (Gonzalez, 2023), math identity (Aguirre et al., 2013), how to overcome math trauma (Vakharia, 2024), how to feel less anxious in general about doing math (Boaler, 2022), and how to help students see the importance of struggle in math (SanGiovanni et al., 2020). What would it take to shift mindsets and beliefs about what it means to be literate to include math—so that once and for all we can dispel the myth that math is optional?

Let me say this up front: Math is *not* optional!

It's time to normalize mathematizing our world. Throughout each day, educators should pause and ask students to use math to solve real problems. "PE starts at 10:15. It is 9:50 now, so how many minutes do we have to finish this activity, clean up, and head to the gym?" Handing out supplies or materials is a great time to consider whether we have enough for everyone to have their own or if we will need to share with a partner or small group. By helping students recognize when math shows up throughout the day, they begin to see that math happens outside of the math block. More importantly, students experience math in practical ways that reinforce the role math plays in helping us to make sense of the world.

> **TIP**
>
> At the start of the day, pose a thought-provoking question that requires math thinking. This can be posted on a board or asked as students trickle into the classroom. For example, "If everyone is present today, I wonder how many (buttons, pockets, etc.) there will be in the classroom. Let's make predictions and see how close we are after taking attendance."

Source: istock.com/lisegagne

APPROACHING THE TEACHING OF MATH LIKE THE TEACHING OF READING AND WRITING

When it is time for reading, students are welcomed to the reading area with smiles, as they anticipate the beautiful pictures and funny voices that will tell the story. They dream of the day they will be fluent readers. When students learn to read and earn the title of Reader, they beam with pride. When learning to read is challenging, students persist because they know they need reading to be successful in life and to pursue their dreams. Before they can read a chapter book independently, they begin to embrace "Reader" as a part of their identity. And if they feel defeated and want to quit reading, we as educators (as well as family and society in general) will not allow them to opt out of improving their reading skills. We exhaust all our resources to get students the support they need to sound out words, comprehend passages, and hopefully develop a love of reading. Ultimately, we push and encourage students to have a healthy relationship with reading and develop the skills needed to become competent readers.

> When learning to read is challenging, students persist because they know they need reading to be successful in life and to pursue their dreams.

The same is true for writing. Even when young children are just scribbling, we encourage them to write their stories and read them to us. They learn that writing is necessary for communication. We can spin tales from our wildest dreams and use our imaginations to create worlds for others to enjoy. When learning the mechanics of writing, when challenged by spelling, or when our sentences run on and on, students are encouraged to "just write." Get your ideas out and organize them in ways that make sense to others later. Making mistakes is an important part of the writing process. Writing a rough draft is expected and revising your thinking is necessary. Whether you are a talented writer or not, "Writer" becomes a part of our academic identity.

Teachers and parents call students little writers before their stories make sense, and children relish the idea that their words have meaning and power. When writing is difficult and the essays are covered in red ink, the suggestions push us to make improvements. There is not an option to give up on writing forever or to choose a college major that will release us from our obligation to write. A student may not have a future as a journalist, but they will learn at least the basic skills needed to become a competent writer.

And then there's math . . .

Many parents and teachers begrudgingly explain the steps needed to solve problems and sigh to ourselves about not being "good at math." Adults unknowingly project negative mindsets and beliefs about mathematics and validate thoughts of perceived inadequacies swirling in children's minds. Parents reassure students by letting them know they were the unlucky recipients of their "non-math gene." Teachers let students know it is okay to not be good at math because they have so many other talents. The problem is that this reinforces the idea that some students can't learn math—which further reinforces that idea that math is optional. (Let's say it again: Math is not optional!)

READ. WRITE. MATH. CONNECT.

We require students to submit rough drafts of their writing and revise based on feedback. Dr. Amanda Jansen makes the case for rough drafts in mathematics to support sense-making or "revising to learn" (Jansen, 2020). Creating a culture in math class where sharing your unfinished ideas is welcomed, is critical to learning and develops collective agency.

HOW DID WE GET HERE?

Historically, the working class was expected to learn only basic computation skills, while thinking deeply about mathematical ideas was considered unnecessary for most students, especially those designated to the laboring class. Education in colonial America was focused on reading and writing, but by the late 19th century business and civic elites shifted the attention to leveraging education to grow the nation's economy. Mathematics became an accelerator for those with "mental superiority," the wealthy, as "proven" by IQ testing, which was used to sort students based on perceived intellectual capacity. Common schools were compared to factories, and the goal was to create punctual, obedient, dependable worker bees, not deep thinkers. Star pupils accepted their positions in the classroom and society, sliding into their "appropriate" jobs designated for those who were not college-bound.

> *Mathematics has a legacy of exclusion because of its use to determine one's station in life.*

Let's be clear: This system was built for the privileged classes, which 100 years ago largely did not include women and girls or people of color. Mathematics has a legacy of exclusion because of its use to determine one's station in life. These discriminatory practices shaped the math landscape and planted the seeds that continue to shape the math identities of many students, especially those from historically marginalized groups.

It is no surprise that the legacy lives on, and students often accept that they are not capable and, in some cases, not worthy of a relationship with mathematics. According to educational psychologists Dale H. Schunk and Barry J. Zimmerman (2008) academic identity comprises self-concept, one's belief about whether they are capable, and self-esteem, whether one believes they are worthy. We have a responsibility to our students to help them develop healthy relationships with mathematics and a positive math identity. Math identity is defined by Aguirre and colleagues (2013) as "the dispositions and deeply held beliefs that students develop about their ability to participate and perform effectively in mathematical contexts and to use mathematics in powerful ways across the contexts of their lives." Our students deserve experiences that will help shape positive math identities because, increasingly, math literacy is required for daily living and for the careers of the future.

Students often dissociate from mathematics because they don't recognize the role math plays in everyday tasks. To counteract this misconception, make time for students to share math happenings regularly. For example, start the day with a simple prompt like, "How did you use math outside of school yesterday?" Initially, you will likely need to provide guidance or give examples from your own life to help them see how math supports us in simple and powerful ways. It is one step toward helping students see that math is for everyone and math is NOT optional.

Source: istock.com/Eleonora_os

The irony is that reading and writing are not intuitive and must be taught, but we are all born with math intuition, the capacity to learn languages, and mathematical ideas. Our brains are wired to recognize patterns and solve problems in different ways. For example, infants as young as 16 weeks (about 3-and-a-half months) old notice the difference in small quantities. Prentice Starkey conducted research with 72 babies at the University of Pennsylvania that supported the claim that babies notice the difference between two versus three black dots on a slide (Starkey & Cooper, 1980). Since this study, similar experiments have been conducted by various research teams, all resulting in the same findings. Each time, babies were able to notice differences in quantities and inconsistencies with objects being switched (Sousa, 2015, p. 11). Number sense is innate, but we typically don't capitalize on this numerosity.

In my time spent working in daycare centers and preschools, I observed little ones bopping to musical beats and mimicking rhythms before they could walk, talk, or learn to count. I've witnessed toddlers sorting shapes and fitting round pegs into round holes based on trial and error. Preschoolers shared snacks and demanded their fair share when their pile of fish crackers was smaller than another's.

One thing we can do to reinforce numerosity and encourage children in math literacy is to attend to the vocabulary with which we describe kids as they engage in math tasks. As children make connections, identify patterns, solve puzzles, or use logical reasoning, what language could we use to support students in recognizing how they are engaging in math tasks outside of math class? What can we call students using mathematics to make sense of the world? Mathers! We are born mathing.

We are born mathing.

Imagine if, on arriving at school, we as educators viewed a child's innate ability and math curiosity as the foundation on which new concepts and deep mathematical understanding could be built. As young students arrive on the first day of school, their mathematical ideas and creative math terminology are celebrated. When they share fantastical stories about imaginary creatures, we capitalize on opportunities to mathematize their stories with them. We present new concepts as problems to solve and discoveries to be made by them, the math explorers who have everything they need to be mathers. We use positive language, share our authentic enthusiasm for even the strategies that are only partially correct and affirm the parts that are mathematically sound before we point out the errors students have made. Students recognize that math is not the enemy, even when they are challenged by it. Students experience math joy and smile often when collaborating with friends to find creative solutions to common problems.

Mathnote
The Math in Music

Jenna Laib's Fraction Intervention

https://qrs.ly/dugnmu0

Did you know that a love of music can be used to connect to fractions? From clapping out rhythms to reading music, fractions help us find the steady beat. Reading music requires a basic understanding of fractions because each measure represents a set length of time. For example, a $\frac{4}{4}$-time signature means a total of four counts for the whole measure. A whole note, two half notes, four quarter notes, or eight eighth notes are equivalent fractions that represent the whole. Check out Jenna Laib's blog post where she shares a Fraction Intervention designed around music.

To nurture positive math identities and present math as useful and enjoyable at any grade level, the adults in the room need to reconcile their differences with mathematics and commit to forging a new path. Math is not new, but new approaches to teaching mathematics are needed. It starts by viewing math as an experience to be had, not just a subject to be taught. We disrupt sense-making when we introduce "the right way" or step-by-step procedures prematurely. Once students believe there is a path the teacher wants them to take, the path less traveled is off the table. It becomes too risky.

Pam Harris's FigureOut Able Math

https://qrs.ly/dkgnmu2

Unfortunately, school is often the place where math dreams come to die. Children are often forced to abandon sense-making in exchange for rules and procedures, and mathematics becomes an unfamiliar thing that is not for everyone. If we want to develop great math thinkers and problem solvers, we need to join Pam Harris in the mission of exchanging algorithms for math flexibility and help students realize that math is "figure-out-able!" (Harris, 2025).

Given our inborn affinity for patterns, shapes, and number sense in general, math more than the other subjects should be at the heart of our academic identities, but it is quite the opposite for many students. They are counting on us to write the counternarrative about what it means to be "good at math." It is time for math dreams to flourish and every student to confidently embrace the idea that they are **Born Mathers!**

Source: Nzingha Montenegro

Let's Redefine the Core Academic Skills!

Based on the Three Rs,

- reading,
- writing, and
- arithmetic,

the core academic skills in school should focus on reading, writing, and basic computation. Consequently, the study of mathematics is often reduced to memorizing facts and procedures to find speedy

> Given our inborn affinity for patterns, shapes, and number sense in general, math more than the other subjects should be at the heart of our academic identities.

CHAPTER 1 • WE ARE ALL READERS, WRITERS, AND MATHERS!

solutions with accuracy. As we discussed earlier, historically, basic computation skills were all that the members of the laboring class needed—at least that was the prevailing wisdom—given mathematics was used as a sorting mechanism. With updated standards and new ideas about what it means to be proficient in mathematics, it is time to change our language, practices, and beliefs. It is time to redefine the core academic skills.

I argue that the core academic skills should be redefined as

- reading
- writing, and
- mathing.

If readers read and writers write, doesn't it make sense that mathers math? Math deserves a verb! A *mather* is defined as a person who uses mathematics to make sense of the world. In case you're wondering, that definition includes everyone. We are all mathers. Introducing this language to our youngest learners is just one way we can ensure they can grow up believing that math is a necessary asset to our identities and pursuit of our dreams. When students engage in tasks in all subjects, we must point out the ways in which math connects. Whether it's a science lesson and students are using math to do conversions, or it's story time and students are counting objects on a page, take time to connect mathematics to the tasks at hand. Children who are exploring their environments with curiosity and thinking mathematically should be celebrated for their math intuition and sense-making of mathematical ideas. It is time to embrace being readers, writers, and mathers.

> Children who are exploring their environments with curiosity and thinking mathematically should be celebrated for their math intuition and sense-making of mathematical ideas.

Source: istock.com/Lordn

Young children with fragile identities need to be acknowledged and affirmed as they develop foundational skills. As young students build towers with blocks in preschool and count using rote memorization of the number sequence, we can encourage them by calling them mathers. Just as we instill the importance of reading and writing, we can include mathing as critical to our future success and connect it to all careers. Children also need to recognize they are mathing when having fun to avoid perpetuating the lie that math is scary and intense.

When children are still learning to read, we encourage them by calling them readers. This includes the student learning to sound out words phonetically and the fluent reader focused on improving their comprehension skills. Young children hold books upside down and pretend to read words, yet they, too, earn the title of Reader. Even when learning to read is challenging, we persist because we know reading is necessary to be successful and pursue our dreams. Learning to read is not presented as optional, and students embrace being a reader as a part of their academic identities, whether they are good at it or not. Even those who have a tumultuous relationship with reading recognize the goal of being a reader as one that cannot be abandoned.

> **TIP**
>
> As often as possible, encourage students by calling them mathers and point out all of the mathing they are doing. Just as we hand out compliments when students are "caught being good," we should cheer our students on—"Wow! Look at all this mathing going on."

We cultivate joy and a love of reading, by reading stories aloud, requiring students to read at home with family members, and allowing students to choose what they read independently. Students choose books based on their interests and often associate quiet reading time with some level of freedom. Reading is positioned as a required skill and something to be enjoyed at your leisure. Math puzzles and games are often reserved as a reward for fast finishers, which reinforces the notion that being fast in math is the goal and only a select few deserve to enjoy math. By having math puzzles and games available outside of the math class and offering fun math activities to all students—even if they don't finish first—just as we offer silent reading time, students begin to connect math with leisure and choice instead of only drill and practice. It is equally important to encourage students to share math activities with their families as it is to encourage reading together.

READ. WRITE. MATH. CONNECT.

When reading books together, pause to ask questions about objects that can be counted, make predictions, and consider how math could be used to solve characters' problems. Any good bedtime story can be mathematized!

See Table 1.1 for some ideas about connecting leisure to math.

1.1 Connecting Leisure to Math

Activity	Cite	Brief Description
Math Puzzles	Math Puzzles https://qrs.ly/9qgnmsn Perplexors: Basic Level | MindWare https://qrs.ly/wxgnmsq Tang Math https://qrs.ly/swgnmss	Assign ungraded math puzzles for "homework" that students can solve with family members or friends. Discuss the experience and solution in class during morning meetings.
Card Games and Board Games	Chutes and Ladders https://qrs.ly/95gnmst Farkle https://qrs.ly/uwgnmsv Prime Climb https://qrs.ly/wsgnmsw Q-Bitz https://qrs.ly/rkgnmsz Shut the Box https://qrs.ly/c4gnmt1 Yahtzee https://qrs.ly/zlgnmt3	Teach students about the math involved in their favorite card games and board games. Many old and new games require a lot of mathing. Let's help children make these connections by calling them out!
Bedtime Math	Bedtime Math https://qrs.ly/ywgnmt4 Bedtime Math Book Series https://qrs.ly/ktgnmt5	Read a story at bedtime and talk through the number relationships. Reading books from the Bedtime Math series or using the app encourages daily engagement with math tasks and exploration.

Mathematics, unlike reading and writing, is treated as a subject that is taken and passed or failed. As students are introduced to new math concepts, the content is taught, assessed, graded, and left in the past. A student who fails the fraction chapter test rarely has the opportunity to learn from the mistakes and deepen their understanding. There isn't a chance to demonstrate the progress that is made with a "do over." The grade is final, and we move on. Even when a student memorizes formulas without understanding and regurgitates them for the test successfully earning an A, they don't retain the information because the brain holds it in temporary storage or short-term memory. When students pass the unit test for geometry, they believe they are done and never have to see that information again. If mathematics was taught as a series of connected ideas that tell a coherent math story, more students could see the benefits of persisting, and failing fractions wouldn't mean failing at mathematics forever. Having a clear purpose for learning the math being taught and being encouraged to solve problems in ways that make sense to them would empower students to take control of their learning.

> **TIP**
>
> When students learn to play math games at school, have them teach the games to family members and friends for "homework." Encourage students to explain to their family and friends how they are using math to win the game.

Knowing that there are practical ways to apply the math learned in classrooms to their lives could lead to a desire to understand math deeply, not just to pass a test. When learning new concepts or skills, math homework should not only be pages of practice. We can design math tasks for homework that inspire students to apply the new skills to practical situations. Knowing that math is used to make sense of the world and solve everyday problems could help students realize that math isn't just for school and using mathematics doesn't end. We must read, write, and math to function in society.

> ## ACTIVITY TO TRY
> *Get Cooking!*
>
> Have students interview a family member about a favorite recipe. They can write the list of ingredients, how much of each ingredient, the steps for preparation, and how many people the dish serves. Students
>
> *(Continued)*

(Continued)

can share pictures of the dish and pictures or video of them preparing it, if appropriate, and discuss any personal connections. Have students work together to determine how to make double or half of the recipe.

Source: istock.com/Prostock-Studio

To support the vision that all students are mathers and that math isn't optional, we need to start distinguishing between using math to prepare for a math or STEM career and using math for life. While it is true that anyone can choose to become a mathematician, it requires a commitment to studying mathematics to the highest levels. Mathematicians study mathematics to solve the problems of the world and sometimes dedicate years to finding solutions to the same problem. Some of the most challenging math problems remain unsolved for many years. This isn't for everyone! Mathematics that is necessary for STEM careers can also be very complex and require deep study of math concepts specific to the chosen field of study. Students need to know that this level of math is within reach but requires hard work, not a special brain. Likewise, all students need to understand that even if they have no plans to be mathematicians or major in a STEM field in college, they still need math as part of their future careers and their lives as good citizens.

Mathnote
Mathers Make History

Calcea Johnson and Ne'Kiya Jackson

https://qrs.ly/bngnmt7

High school students Calcea Johnson and Ne'Kiya Jackson proved a 2,000-year-old math puzzle that was deemed impossible. When it was implied that they were math geniuses or math is just easy for them, they both denied these claims and credited their discoveries to their supportive school environment in which all students are taught that they have immeasurable possibilities and that they can accomplish any of their goals with hard work. At St. Mary's Academy in New Orleans, Louisiana, students believe they are all mathers, even if they choose a career path that is not in the field of STEM.

Unfortunately, in many schools and homes mathematics is often attached to STEM career choices in a way that makes students believe that if they don't choose to pursue these paths they can opt out of mathematics. If we say things like, "If you want to be an engineer, you will need lots of math," a student can easily decide that if they don't choose to be an engineer, they can avoid math forever. Competent mathers can approach high-level math courses with confidence because they know we are all wired to do math and mathers are prepared to work hard to learn new concepts. Dispelling the math gene myth and replacing it with the truth that hard work and perseverance are the criteria for succeeding in math class can begin to shift our mindsets about who belongs in these spaces. (Spoiler alert: *Everyone* belongs!) Ensuring that all students see themselves as mathers provides the counternarrative that mathematics is not just for certain careers, but for everyone. Whether or not they choose to pursue a career in STEM, students (and adults) can maintain a healthy relationship with mathematics and keep on mathing.

> Dispelling the math gene myth and replacing it with the truth that hard work and perseverance are the criteria for succeeding in math class can begin to shift our mindsets about who belongs in these spaces.

So, what does it look like to use mathematics to make sense of the world? It starts by acknowledging how we already use math every day. Adults need to use positive language to describe tasks that involve mathematics instead of reacting to any mention of math with sighs and groans. Children need to see their teachers tackling math tasks with a growth mindset, understanding that abilities can be developed (Dweck, 2016), and viewing challenges as exciting. It is important to expand our definition of mathematics to include simple actions like deciding which route to take home or planning our budget for a trip to the grocery store. We should cheer for children as they play board games and solve puzzles by saying things like, "Look at you mathing!" or "I am so proud of this little mather!"

WHERE'S THE MATH IN THAT?
Musicians Are Mathers!

Some people believe that we have to choose between being creative and being mathematical. The truth is mathers are creative and have a place in both the arts and mathematics. For example, musicians rely on mathematics even when the connection is not made explicitly. Mathematics is all about patterns and so is music. As a matter of fact, mathematics is an integral part of music, and the skills needed to master one enhances the ability to master the other. Both require working memory, sustained focus, practice, curiosity, and collaboration. Both benefit from working independently to master skills, and then coming together to share ideas and create something magical. Reading music requires us to learn new symbols that represent notes, pitches, pace, and patterns. Mathematical symbols demonstrate the relationships between numbers so we can manipulate them to find solutions to all types of problems. Some mathematicians would argue that a well-written proof is a beautiful representation of a puzzle that has been solved. Well, isn't that similar to the combination of notes on sheet music that gives us a beautiful song?

Source: istock.com/nisaul khoiriyah

In an interview with Ashley Cuthbertson, musician, music teacher, and education consultant, we explore the connections between music and mathematics and the impact access to music and math can have on one's identity development.

Deborah: Tell us about your music journey as a learner, teacher, and consultant.

Ashley: I have always been around music. While my family members weren't all musicians, they were all music lovers. I learned to play piano at eight years old, I later learned to play flute, and sang in choruses. Eventually, I trained as an instrumentalist and hoped to be a principal flutist of a major symphony orchestra. I participated in competitions, trained intensely, and even performed at Carnegie Hall. Everything shifted when I decided to teach music on the side to make money while pursuing my music career. I answered an ad to teach music to children and I loved it. Teaching music lit me up in a way that performing never had. I wondered if I should continue on the path of training as a professional musician or switch to pursue a future in education.

(Continued)

(Continued)

> Math is all over music and music is all over math because music is all about patterns.
> —Ashley Cuthbertson

After graduating from college, I studied abroad in Venezuela and volunteered in a program designed to affect social change through music. I saw the power that music had in so many different communities. The work was about so much more than music; it was about children having access to positive experiences and the change that can happen. Music became a vehicle for helping children see the possibilities in life. I knew that this was what I wanted to do. I wanted to see what types of change would be possible through teaching music. Music education could be leveraged to show how we can engage in the world and understand how things work in an accessible way. I entered the teaching force and later earned a Master's degree in Education.

During my 12 years of teaching, I desired to make stronger connections between music and social change. While I didn't know it at the time, I had developed a way of teaching music through a culturally responsive lens. I realized that there was a lot I had to figure out on my own, and that other teachers could benefit from what I had learned. The work I had done in my classroom eventually became the core of my current professional development work. As a consultant, I support educators with examining how music can be a vehicle for change and how students can be engaged in class while making connections to how the world works. I call it culturally responsive music education, and music serves as a vehicle for making connections and granting students access to all types of learning.

Deborah: What was your relationship like with math?

Ashley: In the elementary years, I played the game of school well. But math was the one subject I began to struggle with past middle school. In spite of having great teachers in a supportive school environment, my school math experiences lacked real-world connections. I believe math became hard for me around seventh grade because I was unable to see the relevance. Looking back, I realize that I never understood how learning math could help me understand other things in my life. When I became a teacher, I made it a priority to help students see how music connected to the real life they were experiencing.

Deborah: What connections do you see between music and mathematics?

Ashley: Math is all over music and music is all over math because music is all about patterns. You have to be able to count rhythms and you have to understand the formulas of how music is made up. If you understand the function of a major scale for instance, which is just a formula really, you learn which notes must be included for the scale to function and can replicate it in a different pitch. Another example is when performing musicians need to sight read and recognize patterns that will help them learn a new piece quickly. When musicians are called to sub, they might only have a short period of time to become familiar with the piece. They rely on familiar patterns and formulas from other songs to play new songs fluently.

There are core progressions that our ear naturally hears based on our experiences with basic core music progressions and because our brains are wired to recognize patterns. Pop music is extremely popular because the music relies on the same formulas and chords in different contexts. As a teacher, I looked for real-life connections between music and other subjects, and it was often math connections that were easiest to make. Students learned that if they knew one core progression, they could transfer the knowledge to a different key. If you knew a scale starting on one note on the recorder, you could move your fingers in the same pattern starting on a different note to get similar results.

Deborah: Wow! What an amazing connection between studying patterns to internalize musical formulas and doing the same in mathematics when balancing procedural and conceptual understanding. It aligns with the ongoing debate about rote memorization versus sense-making. The truth is we need both in math and it sounds like we need both in music as well. It is necessary to practice scales and learn to read music as a foundation for being a great musician but there is something more that has to do with intuition and passion. I like to think of it as playing beautiful music requires skill AND heart. Technique leads to correct execution of the notes, but having heart results in a moving performance. I believe this is true in math as well.

What would you tell a student who believes they can be a musician without mathematics?

Ashley: I would start by asking if they want to have a successful career as a musician because that means they need to be business minded. The

(Continued)

(Continued)

> business of being a professional musician is all about math. You have to know how much money you are going to make, how much it costs for the upkeep of your instruments, the type of ongoing training or coaching you will need to pay for, and that is all about math. Ultimately, you will have to plan a budget and determine how many gigs you will need to book to cover all of your expenses.

Deborah: How do we help teachers see the connections between literacy, math, and music?

Ashley: The thing I realized while teaching was that the key to my students being engaged in music was making cross-curricular connections. Arts integration is critical for overall success because in life we don't silo aspects of our lives like we do with subjects in school.

Deborah: What else would you like to share about how your work impacts students' lives and learning?

Ashley: Even though I was a student who loved music and didn't have to be convinced that music was important, as a teacher I encountered students who didn't know why music, art, or even math mattered. We can't just follow the curriculum regardless of what the students are experiencing. The reality is that we work with human beings, and there is not one plan that works for all. You must juxtapose the required standards with what works best for the students in your care. I didn't get into teaching to follow a curriculum. I got into teaching because I saw what was possible when students get involved with music. The most rewarding feedback from educators is when they leave my sessions believing they are now equipped to reach students they've been unable to reach by making content more relevant. When more students have access to high-quality music instruction, they can make connections between themselves and others and the world around them.

In our classrooms, we have many opportunities to invite music into math. We can clap out rhythms as a call and response, when we need to capture students' attention. As students entered my classroom, they heard the soundtracks from movies playing to welcome them. They learned that once the tunes switched to classical music it was time to settle down and get ready to start our day. Together, we can examine song lyrics and poems to discover the mathematical patterns that help them flow. We can take brain breaks with a few rounds of freeze dance, allowing students the chance to get the wiggles out to the music and turn into student statues when the music stops. No matter the ages of children, no matter how cool they pretend to be, a little music goes a long way.

Students who long to be musicians need to understand that they shouldn't run away from math. In truth, if they pursue a musical path, they will

Source: istock.com/FatCamera

be running toward it. They are destined to become pattern seekers and pattern makers, using mathematics to create music for everyone to enjoy. Helping students make connections between math and music is one way to dispel the myth that math is optional and dismantle the binary belief that tells us we must choose because we can't be mathy and musical. You can be in a rock band AND teach math! Just ask Vanessa Vakharia, rock star and math guru (Vakharia, 2024).

1.1 Vanessa Vakharia, lead singer and keytarist of Goodnight Sunrise and author of *Math Therapy*

Source: Allan Fournier

(Continued)

(Continued)

Source: Courtesy of Vanessa Vakharia

1 TIME TO REFLECT AND TAKE ACTION

As educators we have to acknowledge that math is not the problem; our attitudes and negative language contribute to math's bad reputation as being painful and out of reach for most. We need a new approach to teaching math in a new way if we want to rebrand mathematics as something everyone needs and can enjoy. For the next generation to be confident, competent readers, writers, and mathers, they will need to experience a new narrative that depicts the math community of our dreams written collectively by us.

1. What is a belief about mathematics that has been challenged by this chapter?
2. How will you adjust your language and mindset to disrupt the belief that math is optional?
3. Is there work that you need to do to model a growth mindset for your students?
4. How will you intentionally work to build a community of mathers?
5. How can music support you in your mathematical endeavors?

⭐ MATHFIRMATION

Use daily chants and affirmations to support developing math identities. Words are powerful, so use them to help students begin to believe they are mathers.

Say it loud, I'm a mather and I'm proud!

Chants and affirmations can instill positive beliefs and nurture positive math identities in classrooms. When working with third-grade students struggling to memorize their facts, I approached intervention from a different angle. We started and ended every session with positive affirmations. "I have the right to be here, and I was born to do math" and "I am happy. I am good" are two examples of our affirmations. We chanted in a strong voice, a whisper, and in our minds, using hand motions and body language for emphasis. At the end of our time together, 10 intervention sessions, students wrote reflections about their experience. One student wrote, "I know now that I am good, and I am good at math." Associating math struggles with not being good or smart is yet another reason why we need to change our language. Children attend to our words and actions and internalize beliefs based on their perceptions. If we want all students to view themselves as capable mathers, they need to hear us use affirming words to describe their progress.

> **TIP**
>
> Teach students Mathfirmations and encourage them to teach these to at least two other people. Then they can share in class how it felt to make someone else smile or feel proud of being a mather.

2

REDEFINING THE CORE ACADEMIC SKILLS

Source: istock.com/PeopleImages

Bedtime has always been a special time in our family. Whether it was required by the teacher or not, bedtime meant storybooks being read together. It meant bath time, comfy pajamas, and cuddle time. It meant being read to and sometimes taking turns reading one page at a time. It was the moment when we all found out what happens next because last night ended on a cliff-hanger. No matter the grades or classroom experiences, nothing interfered with the joy and enthusiasm shared when it was time to get in a little reading before bed. As the children got older, it meant quiet reading time in pajamas to settle the brain and prepare for a good night's rest. It meant being empowered to choose which book, graphic novel, or magazine to read because it wasn't assigned as homework. Sometimes it meant secretly reading under the covers with a flashlight until you were discovered. Reading before bedtime can set us up for a night filled with magical dreams or simply help us doze off as we try to reach a good stopping place in the story.

Many families have bedtime routines that include story time, so why is it equally as common that families have never considered what a math routine at bedtime might look like? Why isn't there any consideration for establishing routines for family math time over the course of the evening? Why do we accept that reading deserves daily practices and bonding experiences, while math is done at the counter or dinner table in isolation? How did reading become warm and fuzzy while math gained the reputation of being stiff with sharp edges? More importantly, what can be done to shift mindsets and beliefs around how math can be enjoyed by all?

Reading has always been promoted as **fun**damental and necessary for life. Students are encouraged to develop the habits of good readers, and teachers work to nurture a love of books for even the most reluctant students. There has never been a narrative about reading that would make anyone believe that we could hop off the reading train if the ride becomes a little bumpy. In most teacher training programs, there are several required courses focused on the developmental stages of reading and additional core courses and electives designed to hone in on teaching practices to support high-quality literacy instruction. Professional learning opportunities are offered throughout one's teaching career to infuse the current research on reading and writing. Whether or not everyone agrees on the latest philosophy, there are passionate educators advocating for literacy instruction to be up to par so that all students learn to read. The love of reading is a gift, but the skill of reading is a mandate.

When exploring ways to integrate across content areas, it goes without saying that reading and writing will be a part of the plan. Using nonfiction texts to support social studies or historical fiction to inspire writing about a specific time period in history are common practices. Students follow templates for writing up results of their science

> *Mathematics is taught in a silo, and students are often alarmed when words show up in math class.*

observations and map out strategies for their design thinking projects. Reading and writing are considered no matter the subject, likely because they are viewed as core academic skills. But shouldn't mathematics be treated the same way? We limit opportunities to mathematize the work students are doing in all subjects and we minimize the impact these experiences can have on building positive math identities and ensuring long-term academic success. Mathematics is taught in a silo, and students are often alarmed when words show up in math class. Word problems cause anxiety for many students and adults because of this disconnect and the missed opportunity to leverage what students enjoy about reading to support mathematical understanding. Should reading instruction be prioritized over math instruction in the early grades? Research tells us that it should not.

Early Literacy + Early Numeracy = Academic Success

What if I told you that early numeracy skills are one of the best predictors of long-term academic success, not only in math but across other content areas as well? In a study initially conducted to determine the connection between school readiness and later academic achievement, Duncan and colleagues (2007), discovered that early math skills were the greatest predictor of academic success, followed by reading and attention skills. Yes, you heard right, early mathematics skills—not reading—was found to best predict future reading and math achievement. While reading and language development were important, foundational math skills were found to matter most. Playful math experiences and leveraging children's born mather abilities, such as noticing patterns and manipulating numbers, are critical to laying the groundwork for students to thrive in later grades.

A similar study conducted by Romano and others (2010), looked at family relationships and support, social-emotional behaviors, and reading and math abilities of kindergarten students. They found that early literacy and numeracy skills were the strongest predictor of academic achievement through third grade.

> **TIP**
>
> Simple adjustments to language can make a big difference in students' attitudes toward tasks. The next time you are solving word problems with students, introduce the "math story" with a smile and curiosity about the problem you will **get** to solve.

In a more recent study, Mononen and colleagues (2014) found that later math achievement could be predicted by children's performance on preschool math tasks. They also found that supporting students early on with moving through the concrete-representational-abstract process and engaging them in games improved the performance of those struggling with math tasks. This means that early interventions in mathematics have a lasting impact on students' academic achievement overall.

Source: istock.com/FatCamera

So, do I have your attention now?

Shouldn't we invest more time, energy, and resources in professional learning focused on mathematics instruction? Isn't it critically important for educators to prioritize building a community of readers, writers, and **mathers**? What can we learn from literacy instruction to support us as we strive to redefine the core academic skills as reading, writing, and mathing? The first lesson is that when we introduce reading and writing, we take the time to lay a solid foundation. We spend time learning about the sounds 'a' makes, the symbol that represents it, and how it can help us form words. Children learn one letter at a time, blend letters, and eventually decode words. They are read to before they can read their own stories and

Early mathematics skills—not reading—was found to best predict future reading and math achievement.

What can we learn from literacy instruction to support us as we strive to redefine the core academic skills as reading, writing, and mathing?

are given opportunities to discuss elements of a story they cannot decode independently yet. Why do we take this approach? Because we want students to learn to read fluently and accurately, but we ultimately want them to comprehend what they are reading. This takes time. So, what does this look like in math?

Well, let's start with the basics. Just as reading begins with A, B, C, mathing begins with 1, 2, 3. We want students to be fluent and accurate, but we also want them to develop conceptual understanding. We want them to be flexible thinkers who understand number relationships, and that takes time.

READ. WRITE. MATH. CONNECT.

When teaching students to read, we introduce letters and their sounds one at a time. Students explore the things in their environment that start with that letter and practice making the letter sounds before blending the letters to make words. Consider introducing numerals in similar ways. Before students are introduced to expressions and equations, they need lots of practice connecting the number symbols to quantities. As they learn to write each numeral, take time to support them with connecting the symbol to multiple representations like fingers, drawings, and items in their environment. Answering the question "How many?" is critical to developing number sense.

... Or Is It?

Parents of kindergarteners often complain, "Too much time is spent just counting!" Many teachers believe advancing students to writing expressions and equations is the solution. Students' use of symbolic representations of numbers without conceptual understanding will not result in mathematical prowess. Instead, students recite fact families without understanding the relationship between addition and subtraction and without learning the true meaning of the equal sign.

A weak foundation in preschool and kindergarten can often lead to a bumpy math journey and disrupt positive math identity development. If we spend enough time early on solidifying counting strategies, students won't get stuck there. Eventually, they will make the connections needed to shift toward additive and multiplicative reasoning (Harris, 2025).

Sometimes, the "struggles" detected in the upper elementary grades are directly connected to a lack of number sense and flexibility with numbers. Children develop number sense by using mathematics to make sense of their environments even before they realize they are doing it. This sense-making lays the foundation for an understanding of number relationships, flexibility with numbers, and inventive strategies for solving problems. One way we can support students in the early grades is to spend a LOT of time counting.

Since I am not currently in the classroom, I don't often get to work with the littles, so I hesitated to accept a kindergarten student for virtual math coaching. I made an exception for a delightful little lady named Zoë. She shared that she did not like math and wasn't thrilled that we would be working on math every time we met. Watching her develop as a doer of mathematics, particularly with counting, was rewarding and enlightening. While we explored many concepts, and Zoë blossomed in many ways, let's zoom in on the unfolding of this little mather's growth, specifically with counting.

> A weak foundation in preschool and kindergarten can often lead to a bumpy math journey and disrupt positive math identity development.

TIP

At the start of the day, introduce something the students will count as the day goes on. For example, every time we hear a bell, or every time we see a red shirt, etc. we will add it to our chart. You could also prepare a chart in advance with images that students will look for in the classroom to count throughout the day, like a scavenger hunt.

Five Counting Principles by Zak Champagne and Rob Schoen

https://qrs.ly/11gnmu7

https://teachingisproblemsolving.org

Let's examine the complexity of counting by walking through our school year journey of the Common Core State Standard K.CC.B.4 (Counting to tell the number of objects) and the exploration of the five counting principles outlined by Champagne and Schoen (2020).

1. Standard Order Principle

Knowing the number sequence and understanding that it matters.

From our first session, it was clear that Zoë knew and understood standard order. She counted to 120 without hesitation and only stopped because I told her she could. So, what did this mean for our work together? It meant that we continued to practice counting every time we met. She counted up to a given number, which required her to know when to stop. This seems like no big deal, but children who "sing" their numbers often keep going even when given a target number. Zoë started the year only counting by 1, but by the spring she was counting by 10, and working on counting by 5. When I asked her how high she thought she could count, she confidently told me she could count to 1,000, but it would take too long. Instead, she decided to count to 200, and she did. And guess what? She was not bored with counting, nor did she believe that "counting was for babies." As a matter of fact, smiles were often involved. More importantly, Zoë was not only a rote counter, but she also understood conceptually that the number order matters and represents quantities.

2. One-to-One Correspondence Principle

Pairing one number name to one object as you count.

Zoë consistently demonstrated one-to-one correspondence when counting objects, and she also double checks, self-corrects, and adjusts when she makes a mistake. We counted objects every time we met. From the beginning, she organized the objects while counting, but with larger numbers, she sometimes ran out of space. This counting principle takes time to develop; some children appear to "have it" one day and lose it the next. Guess what? Children need to practice counting objects. They should count anything and everything, and they should count often. Children should always count objects; creating counting collections can keep them interested.

There were times when I tried to nudge Zoë to count based on her groupings, especially when she had lost her place, but she would start over and count by 1, tagging each object. As her organization became more sophisticated, she did not have to "count all" if she lost her place. Eventually, Zoë didn't have to physically touch each object, she could count "in her head" while nodding along. As seen in the next photo, she learned to model her thinking by creating towers of ten when using cubes, counted by 10, and then counted on by 1 to reach a target number like 67. It is important to note that progress happened over time and not before she was ready.

3. Cardinality Principle

Understanding the number of elements that are in a set and understanding that the last number said when counting is the value of the set.

4. Conservation of Cardinality Principle

Understanding that even when the order or arrangement of the set of objects changes, if none were added or taken away the total remains the same.

Zoë was consistently showing evidence of cardinality long before conservation of cardinality. There were often times she recounted objects to make sure it was still the same amount, even though she had just counted it. As conservation of cardinality developed, it filled me with delight to watch her reaction when I asked, "How many are there now?" and she looked perplexed, maybe even offended. She boldly stated some version of, "It's still 13. I just counted it."

Many students can count objects and appear to understand cardinality. It is the ability to answer the question, "How many are there?" that provides evidence of true understanding. Many children can count a set of objects without recognizing that the last number stated represents the total number of objects in the set. When students are asked, "How many is that?" and they start counting the objects again, cardinality is not fully developed. Even when students understand cardinality, the ability to conserve cardinality may still be absent.

> **ACTIVITY TO TRY**
>
> Have students count objects in different ways. Give the student the number of objects that is one more than the amount they can count easily. Have them count the objects that are on the table in a pile, in a line, and scattered. Then have them count the objects as they place them into a box one at a time. Then have them count the objects as they take them out of the box and place them on the table. Each time ask, "How many?" For example, "How many objects are on the table?" "How many objects are inside the box?" or "How many did you take out of the box?" You can provide a sentence frame for language support.

Splat! Steve Wyborney's Blog

https://qrs.ly/3lgnmtb

5. The Successor Principle

Understanding that there is a number that follows each number that is 1 more.

Zoë showed a lot of growth during her kindergarten year, as she practiced counting not only objects, but images on a screen. Sometimes the shift happened so quickly! One week she was tagging every virtual object using annotate tools and counting aloud, and the next week she was counting silently stating only the total number of images. During one session when working on a Splat by Steve Wyborney, Zoë surprised herself when she decided not to use blocks to help her think about the total number of dots on the screen. She said, "I don't think I'm going to build it." Then she proceeded to think aloud counting on 1 by 1 to get from 15 to 18. It was a proud moment for her, and after that she counted on and counted back to find the rest of the solutions. Just like that, she recognized that she understood something new mathematically that would always work.

Students demonstrate this principle when they know the value of a set and can tell you how much would be in the set if one more was added to it. They also show evidence of this understanding when they count on instead of counting all. By the end of kindergarten, Zoë understood the successor principle and demonstrated conceptual understanding.

It is safe to say that Zoë has demonstrated evidence of conceptual understanding of all of the counting principles, but it took time to develop. Sometimes, I wanted to hurry her along because she seemed ready, but stepping back instead was the right move. Our time together was not to ensure that Zoë entered first grade knowing all of her addition and subtraction facts, even though she could quickly find solutions using varied strategies. The real goal was to support the development of a positive math identity. Zoë was beyond ready for first-grade mathematics, but more importantly she believed that she is a mather, a problem solver, and a doer and creator of mathematics. All the parts of the standard were met, while enjoying math. If students experience joy in math class, they will do the work necessary to achieve success. (See Figure 2.1.)

2.1 Zoë the Mather in Second Grade

Source: Vanity Jenkins

Why does this matter? It matters because counting is NOT as easy as 1-2-3, so we must give students the time and space they need to build a solid foundation. If we don't, the house we build is destined to crumble. Too often we aren't giving students enough opportunities to count, especially beyond first grade. Students need to count beyond 120. They need to count alone, with a partner, and sometimes as a whole group in a choral count. By hearing others count fluently, their counting skills can improve. Students need to count forward and backward. They need to count by 2, 5, 10, and 100, but they also need to count by $\frac{1}{2}$, $\frac{1}{4}$, and $\frac{1}{3}$. Students need to skip count by 3, not starting with three, but counting on from any number. We run the risk of teaching only sing-song rote counting if we don't explore options for students to play with counting. The patterns they discover during choral counts can help them deeply understand number relationships in ways we cannot explicitly teach.

Counting collections are another way to support students with developing strategies for grouping and counting, and to think about finding totals of quantities using additive and multiplicative reasoning. In *Choral Counting &*

> Counting is NOT as easy as 1-2-3, so we must give students the time and space they need to build a solid foundation.

CHAPTER 2 • REDEFINING THE CORE ACADEMIC SKILLS

Counting Collections (2018), Franke and colleagues emphasize that counting establishes a core foundation for understanding place value, the composition and decomposition of numbers, and number relationships. The authors advocate for counting everything all the time, well into the upper elementary grades and share activities and guidance for implementing choral counting and counting collections in pre-K through fifth grade.

Source: istock.com/BlessedSelections

My colleagues and I saw the benefits of counting collections when working with third and fifth graders enrolled in the **Trust the Mather in Me** summer program. We gave students a task that required them to work in groups to count hundreds of beads, dice, and other items. The students were fully invested because we shared that we needed their help buying supplies for the community STEAM event we were hosting at the end of camp. Each group estimated how many items they had, counted to find the actual quantities, and then figured out how many more items were needed to meet their target number. Productive struggle had entered the building.

We witnessed students counting by 1, getting frustrated by losing their place in the count, and adopting a new strategy based on observations or recommendations. We watched students strategize about how they could divide and conquer, honoring one another's counting strategies, and then working together to find the combined total. We heard laughter and saw smiles on the faces of students who earlier proclaimed the task might be impossible. We encouraged, questioned, nudged, and celebrated small victories. In the end, every group was successful, and every student stood a little taller that morning. These students learned to trust the mather in themselves to solve a real-world problem.

> **TIP**
>
> Counting Collections are great for "homework." Use plastic baggies to send home objects for students to count and record how many they counted. You can also send home an empty bag with a target number for students to make a collection to bring to share with the class.

The truth is, we didn't need to order supplies, but our "plight" made the task relevant to the students. The key is that there was a community STEAM event planned, and we did need these supplies, but everything had been ordered in advance. Sometimes we count for the practice of counting, but sometimes we need a purpose to be motivated and to persist through the struggle. In week one of a two-week program with students who barely knew us, we opted for the inspiration of a real-world problem.

Check out the following activities for some suggestions to get students counting!

ACTIVITIES TO TRY TO GET STUDENTS COUNTING

Activity/Resource

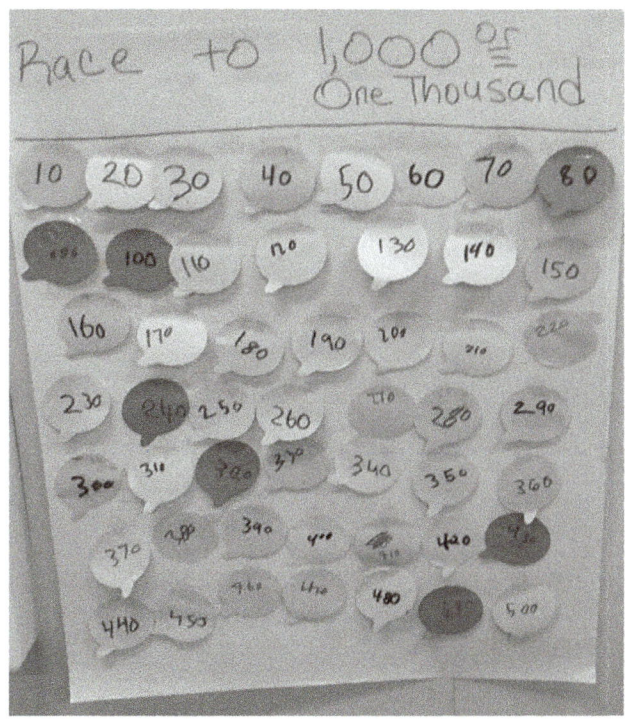

CHAPTER 2 • REDEFINING THE CORE ACADEMIC SKILLS

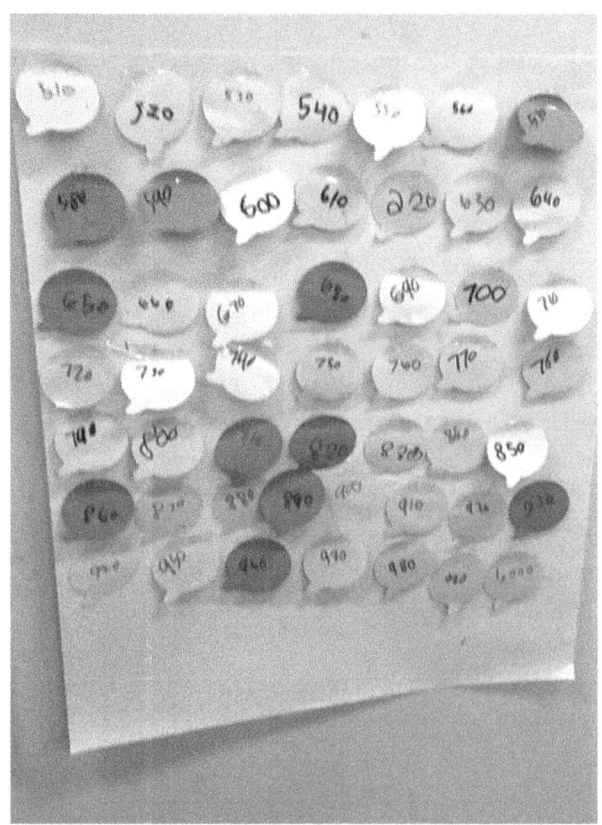

Cite/Example

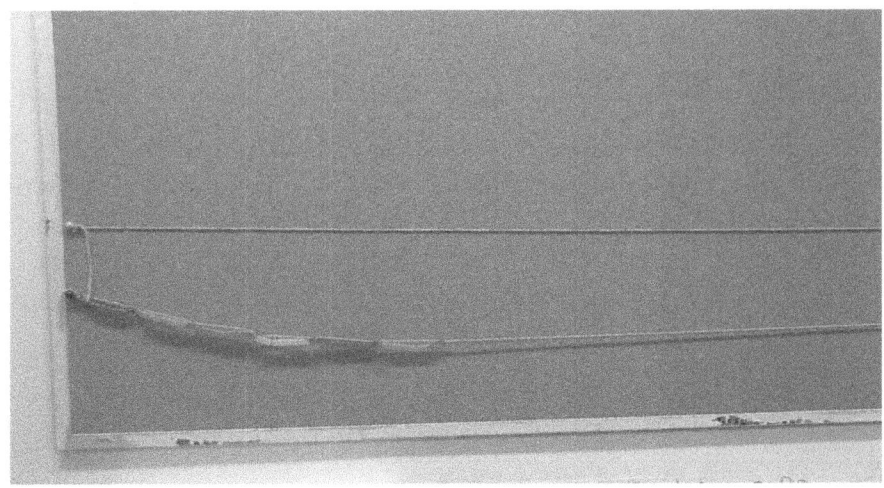

Brief Description

Students count beads and string them on large paper clips in groups of 10. As they turn them into the trays, the teacher strings them onto the board. Students then write the next number they would say when counting to 1,000 by 10. The visual of 1,000 beads is important so students see what the quantity looks like. Separate each 100 with a marker so students can clearly see 10 hundreds represented.

Activity/Resource

Choral Counting

Counting Collections

Cite/Example

Choral Counting & Counting Collections: Transforming the PreK-5 Math Classroom (Franke et al., 2018)

Brief Description

Using choral counts and counting collections consistently engages students in activities that are fun and foster learning. Students explore number patterns, place value, and number operations in a collective learning experience.

Activity/Resource

Splat

Cite/Example

Splat!—Steve Wyborney's Blog (Wyborney, 2017)

Brief Description

Using quick images, students get practice with subitizing and part-whole relationships. By showing all of the dots and then covering part of the group, you can extend this warmup activity in varied ways. These experiences can build an understanding of the relationship between adding and subtracting as the quantities get larger.

Mathnote
A Baker's Dozen

Did you know that a baker's dozen is 13 not because bakers in medieval England couldn't count, but because they wanted to avoid a beating? This is one theory based on laws written to keep bakers from selling loaves that were too light. The cost of the wheat used and the cost of the loaf were supposed to be about the same. Since baking isn't an exact science, bakers sometimes threw in an extra loaf to ensure the weight of a dozen loaves measured up. They didn't want to be fined and they certainly didn't want to be flogged for coming up short.

How can we leverage what we know about teaching literacy and our beliefs about good readers to build a community of mathers? It is a lot easier than we think. The first step is to consistently include mathematics as one of the core academic skills in our words and deeds. Intentionally demonstrate the cross-curricular connections, weave mathematics throughout all content areas, and highlight the mathing that is going on all day in our classrooms and beyond. When students begin to realize that math resides outside of the math block and can be used to make sense of the world, just like reading and writing, they will learn to embrace mather as a part of their academic identities. Mathematics is meant to be a sense-making super power, but too often it is viewed as Kryptonite. In other words, students need to mathematize EVERYTHING!

For decades, in classrooms far and wide, we have seen posters outlining the habits of good readers. These posters are hanging as reminders to students, no matter their relationship with reading or how far along they are in the process of learning, to set aspirational goals. What is interesting is I cannot remember ever seeing a similar poster outlining the good habits of mathers. Of course, this idea of mathers is new, but hopefully we can learn what it might look like by checking out these not so exclusive reading strategies.

So, what are the habits of mind that students are taught as the habits of good readers?

Good readers . . .

1. use the information they know to help them understand new information in the text.

2. decide which details are important and what information is not needed for their understanding.

3. ask lots of questions in their minds and of others to gain clarity and focus their ideas.

4. visualize by creating images or "making a movie in their minds" to help them imagine what the sensory experience might feel like as they read.

5. make inferences using their prior knowledge and information from the text to interpret, predict, and draw conclusions about the text.

6. synthesize information by making connections across old and new ideas, filtering out obsolete understandings, and creating new learning pathways.

7. use varied strategies to make sense of information they do not comprehend like rereading, comparing and contrasting ideas, considering the context, or skipping ahead and coming back.

And, what would it take to adapt these practices to work for mathers?

Good mathers . . .

1. use the information they know to help them understand new information in **the task**.

2. decide which details are important and what information is not needed for their understanding.

3. ask lots of questions in their minds and of others to gain clarity and focus their ideas.

4. visualize by **creating pictures, diagrams, or models** to help them imagine what the **solution path might be**.

5. make inferences using their prior knowledge and information from **the task** to interpret, **estimate, strategize**, and draw conclusions about **the task**.

6. synthesize information by making connections **across strategies, considering number relationships**, filtering out obsolete understandings, and creating new learning pathways.

7. use varied strategies to make sense of information they do not comprehend like rereading, comparing and contrasting **mathematical ideas**, considering the context, or skipping ahead and coming back.

What do you notice? What do you wonder?

I notice there are more similarities than differences, so we should consider sharing the *Habits of Good Learners* instead of treating these subjects so differently. We can leverage what we know about supporting students with developing good reading habits to help them realize they need very similar habits to be good mathers. The focus here is on establishing a norm that ALL students need to develop these habits to learn. Students will recognize that if they CAN read, they are readers, if they CAN write, they are writers, and if they CAN math, they are mathers. This distinction is important because they don't have to **love** all subjects to be successful in them. We want students to embrace reader, writer, and mather as a part of their academic identities. Later, they can decide if they will invest the time and practice to excel in these areas or choose career paths that require them to become more than proficient. Our job is to ensure that they are equipped to make these decisions from a place of confidence not fear. The way we do this is by cultivating a love of learning, a passion for problem-solving, and a mindset that anything is possible.

> Students don't have to **love** all subjects to be successful in them.

Source: istock.com/Jacob Wackerhausen

READ. WRITE. MATH. CONNECT.

When assigning reading for homework, give options to families. For some students, reading for 20 minutes will work just fine. Other students who are easily distracted or avoid reading practice often benefit from being assigned a number of pages or chapters. This way more time is spent on task instead of taking trips to the restroom, getting water, or asking if the time is up yet. It also gives children a little more control over how long they read with a concrete goal instead of the abstraction of time.

WHERE'S THE MATH IN THAT?
Bakers Are Mathers!

Whenever I entered Sara Fludd's Pop Goes the Waffle Café, the sweet and savory aromas made my mouth water. I was honored to look up at the menu and see The Deb, an item added to the menu and named after me because I ordered it every time I stopped by, even though it wasn't an option. This was one of the benefits of knowing the owner of this fine establishment. It was a delicious waffle with sticks of chocolate in the center, topped with fresh strawberries and whipped cream.

Families came in to enjoy Liege waffles with sugar pearls baked inside, topped with fresh fruits or Fruity Pebbles if they wanted something extra sweet. They could also order a breakfast sandwich wrapped in thin waffles as the "bread." During festivals, you could get a waffle you could walk with, a waffle pop drizzled with chocolate syrup and cookie crumbles. Another favorite of mine was the shrimp and grits waffle, a waffle made from cheese grits topped with shrimp in a sauce that was simply divine. The options were endless and the menu changed often to match your favorite seasonal treats or to take advantage of the fruits in season. You name it, you could have it. You could even get waffles to go when you ordered individually wrapped waffle donuts or when you purchased waffle packs so you could freeze a stack and toast them later. Pop Goes the Waffle was thriving and serving the community, but the owner claimed that she was absolutely NOT a math person.

Deborah: Tell us about your baking journey.

Sara: Well, I grew up in the south with strong Black women who cooked and baked everything from scratch. On a regular day we had pancakes, grits, eggs, and biscuits. Our family grew our own vegetables and baked our own bread. I was often the little helper, so I learned a lot from these women. My grandmother was recognized as the matriarch and baker, so I definitely inherited that gift from her and learned at her side. Recipes were not written down, so some things I have made from memory, while others have been lost. When I close my eyes,

I can still smell some of my grandma's cakes fresh out of the oven. Over the years, I have been inspired by these memories and created recipes of my own.

When I became a mom, I knew I wanted weekends and holidays to be special and the best way I knew to do that was with food. We had homemade biscuits, pancakes, cinnamon rolls, and anything my daughter requested. I loved a good challenge, so I started finding recipes to try and often made them better. Sometimes we had the pleasure of entertaining friends and family, and they were always amazed by the southern hospitality and home cooked meals, especially the fabulous desserts.

Some of my favorite things to bake are layered cakes. I am known for my red velvet, carrot cake, and thin layered caramel cake. My grandmother's caramel cake recipe with a few adjustments of my own has made many of my friends break a diet or two. Eventually, I perfected a caramel cupcake recipe, which was easier to share. Mostly, I baked because I enjoyed making people smile and seeing them close their eyes to savor the flavorful explosion in their mouths. I didn't bake for the compliments. It was a rewarding experience when you've created something that gets that kind of reaction from people.

Deborah: So how did waffles get into the picture?

Sara: When I was little, I had pancakes all the time, but I had never had waffles. I remember going to the grocery store with my mom and seeing Eggo Waffles. I was mesmerized and begged to try them. I am not sure why, but my mom gave in and bought a box. I fell in love.

It wasn't until my daughter was older that I tried making waffles from scratch. For Mother's Day, I received a waffle iron as a gift, so I decided I would experiment. We had moved to the northeast, so I thought it might make a great snow day tradition to have waffles for breakfast. That winter we had so many snow days that I started getting really creative. First, we experimented with different toppings and fillings, but eventually we tried a few savory things like eggs and grits. That winter, we waffled everything! All of our attempts were delicious, and we joked about one day opening a waffle shop. When the time came, my husband suggested the name Pop Goes the Waffle.

When we moved to Florida, we started with a tent at local markets before we invested in a food truck. The waffles became so popular that we eventually opened a café. The café was a dream come true because it felt like coming home. In addition to waffles, we sold baked goods like olive oil loaves, banana

(Continued)

(Continued)

bread, and waffle donuts with all kinds of dips and toppings. We also offered breakfast items reminiscent of my South Carolina breakfast table. When we hosted brunches, the menu included biscuits, hash browns, and grits, all made in the waffle iron. Our breakfast sandwiches wrapped in a special thin waffle were also a big hit.

2.2 The Queen Waffler Serving Up Waffles

Source: Boyzell Hosey

Deborah: What was your relationship like with math?

Sara: In the elementary years, math was okay but it sometimes took me a little longer to understand what the teacher was explaining. My mom was a well-known teacher in the small-town community so there was a lot of pressure to be an excellent student. I worked hard to just do okay in math, so I didn't like it much at all. I LOVED reading and writing and excelled in both, so I just figured I wasn't a math person.

When I got to middle school, math officially became my worst enemy. Sometimes I think I passed classes because my teachers respected my mom too much to fail me. As soon as I could, I opted out of math and chose a major in college that required very little mathematics. As an English major with a minor in journalism, mathematics was in my rearview and I was thrilled.

I am not sure if my struggles with math were legitimate or if I simply gave up, but as an adult I cringed at the thought of taking on math tasks. I refused to help my daughter with math homework as early as second grade because I was nervous I would "mess her up." Luckily, my husband was a math teacher, so I was off the hook.

Deborah: What connections do you see between baking and mathematics?

Sara: The funny thing is I never thought about math being a part of baking. It wasn't until a friend pointed it out to me that I realized how much math I was doing without thinking about it. When I am trying out a new recipe, I play around with the ingredients to find balance and get the flavor I want. I had NO idea that it was connected to proportional reasoning or that it was considered mathematics. When I needed to double or triple recipes for large parties, it wasn't always easy to get it perfect because you can't simply double all of the ingredients. It requires what I now know is called irregular scaling. It usually meant doubling base ingredients, but scaling down for the spices or flavor elements.

Of course, I knew the measurements were mathematical, but for some reason I wasn't intimidated by converting fractions when it was to get my recipe just right. It seemed like this was a different kind of math because I believed that I was terrible at math but this came easy to me. When I became a business owner, I had to admit that even though learning math was challenging, I wasn't actually bad at math. I think I held on to the trauma of being embarrassed when put on the spot to say my times tables or for asking questions that "I should've known the answer to." I'm pretty sure that I was so scared to get things wrong that after a while I just gave up.

Now on a regular day I am thinking about payroll, inventory, prepping for orders, and scaling my business. And preparing the waffles requires a great deal of mathematics! I measure out every ingredient meticulously; roll out the dough balls, which rise for an exact amount of time; and organize them on trays. The waffle irons need to be at a precise temperature. I have timers set to ensure they are the perfect shade of brown. And I arrange the waffles on the trays in arrays of the same amount each time to make it easy to keep track of how many are made and how many more need to be made just by looking at the full trays. This helps me with keeping track of orders that need to be filled. So, I guess there are many connections to mathematics.

Deborah: Well, you are a Mather for sure! What would you suggest to teachers who want to help students see connections between mathematics and baking?

Sara: I think the most important thing is to help students know when they are using math. I was using measuring tools to help my grandmothers and mother before I could reach the counter,

(Continued)

(Continued)

but I didn't see that as math. I could calculate mentally at home when thinking about how much we needed to make or buy based on who was coming to dinner or for a large family gathering, but I froze when asked to recite my facts. Grocery shopping with my mom was one of my favorite things to do, which required planning a budget and keeping track of our spending while we shopped. Again, I had no idea how much math I was doing every day and believed that I was terrible at math well into adulthood.

I guess my advice isn't specific to baking, but I hope teachers help students realize the fun ways they are engaging in math tasks inside and outside of school. My fear of mathematics could've kept me from pursuing my dream of opening a waffle shop, but I am thrilled that it didn't. My husband encouraged me and pointed out to me that I didn't need his help as much as I thought I did. He helped me realize I was more than capable of solving math problems on my own. If teachers do this for students, I believe math won't interfere with pursuing their passions.

Even though we closed the café after the challenges the pandemic caused with the supply chain, I am glad I had the chance to bring a little waffly goodness to folks all across the country.

Baking together is something that families could use to have math talks with their children and yummy moments connected to math. This means that cooking together can be a fun bonding activity and family math time. Just as we orchestrate special reading practices for winding down every evening, math can be included regularly in our homes. A wonderful first step would be to help families recognize how they are already mathing and encourage them to include their children in these activities. Even when performing chores, there is an opportunity to estimate how long it will take and, more importantly, how close a child might be to the free time that comes when chores are done. As much as possible, let's find positive ways to reinforce the use of mathematics to make sense of the world inside and outside of the classroom. Maybe as educators we will reap the benefits of tasty baked treats from families who are practicing math together!

TIME TO REFLECT AND TAKE ACTION   2

For many years, reading and writing have been viewed as the most important core subjects for students, but there is so much evidence that supports a different narrative. Reading, writing, and arithmetic implied that all students needed was to be literate and have basic computation skills. This was because years ago, our society needed human computers who could find calculations quickly and accurately, not deep mathematical thinkers. Now that we have phones that are computers, it is safe to say that we don't need to focus only on basic math skills.

It would be great if we updated the definition of literacy to include numeracy, especially since early numeracy has been proven to be critical to academic success. Whether or not that happens, we have a responsibility as educators to lead the way. Let's cultivate learning spaces that celebrate mathematics with equal fervor as we do reading and writing and share our enthusiasm with the community at large.

1. What is one way you can support families with creating math routines at home?
2. How will you adjust your language and mindset to help students recognize the math that is connected to things they enjoy?
3. What practices will you try at home and share with your students?
4. How do you plan to highlight mathematics across other content areas?

MATHFIRMATION

It can help students get excited when we put our mathy lyrics to the tune of popular songs. Using the call and response structure and building on the lyrics from the hook of the song *I Can* by Nas, we had a daily chant in the two-week Trust the Mather in Me summer program. Our third- and fifth-grade students battled it out to see who did it best by the last day of camp.

I Can by **Nas**
https://qrs.ly/ifgnmte

(Continued)

(Continued)

 Dionne Aminata, CEO Math Trust, and Deborah Peart Crayton leading the Trust the Mather in Me Summer Math Camp

Source: Deborah Peart Crayton

> I know I can, I know I can
>
> Be what I wc
>
> If I work har
>
> I'll be where
>
> I know I can,
>
> Be what I wc
>
> I can read, w
>
> Trust the Ma

3

LET'S SHARE STORIES NOT PROBLEMS!

Source: istock.com/Wavebreakmedia

If we could read the minds of children in classrooms all around the country, we might hear, "Oh NO! What are words doing in math class?! Math is supposed to be about numbers not words. What will I do? Yes, I remember. CUBES! First, I need to **C**ircle the numbers. Then, I **U**nderline the question and **B**ox the key words. Next, I **E**valuate and write an equation. Finally, I **S**olve and check. Or is the **E** for eliminate extra details? How do I know which words are key? How do I know where to begin? Ugh, I hate word problems."

C	Circle the Key Numbers
U	Underline the Question
B	Box the Key Words
E	Eliminate Extra Details
S	Solve the Problem

Nowhere in this prescribed list is there guidance about reading the problem and making sense of the context. As students learn to rely on a mnemonic to remember the steps to solve word problems, we need to ask ourselves if this is a strategy we would offer when students approach unfamiliar text during the literacy block.

Spoiler alert: It is not.

Readers read to gather information, to understand new ideas, for entertainment, and to find answers to questions. So, why is it common to hear educators offering advice that instructs students to abandon their reading strategies and ignore comprehension when it is time to solve word problems? The eye rolls and unspoken words of students and teachers imply that contexts disrupt math class. In reality, the opposite is true. Contexts illuminate mathematical ideas and deepen understanding. So, how do we leverage the lessons we've learned in literacy? We focus on **sharing stories** in math class instead of **assigning** word **problems**.

We teach the elements of a story to help students follow the rising action and draw conclusions based on the arc of the story and character development. Students learn that every story has a setting that could be the time of day, the location, or the time period in history. Every story has a cast of characters who are relating to one another. There is a protagonist or "hero" and the antagonist whose job is to disrupt the hero's journey. The job of the reader is to follow the story line, which includes actions, events, decisions to be made, and problems to solve. We accompany the characters on an adventure, an emotional trek, or a mysterious mission, and as we approach the denouement, we read feverishly to find resolve as the story ends with the problem solved. Children learn that stories are for entertainment, and we go along for the ride to see where the author's imagination will take us.

What if I told you that students could experience something similar in math class if we shared math stories instead of word problems? And one of the cool things about math stories is that the reader GETS to solve the problem and save the day.

You're probably thinking, "But a word problem is often just a few sentences. It hardly qualifies as a story." Well, allow me to introduce you to the micro-story. "What is a micro-story?" you might ask. A micro-story is an extremely short narrative of 300 words or less. When writing micro-stories, authors limit the number of characters and focus on one subject or small idea. Sometimes micro-stories are accompanied by a powerful image. Another key feature of a micro-story is that in very few words, the author uses descriptive language to tap into the reader's emotions and describe a problem that matters.

The essence of a micro-story is "Show Don't Tell." The writer evokes curiosity by describing situations and using words to paint a picture for the reader. Math stories need to be short but also relatable. So not only should we be sharing math stories with our students, but we should also be writing them because that is the best way to ensure they are relevant. If we invest time to get to know our students, we can author math stories that engage them and sustain their interest. Students invested in the situation and the problem that needs to be solved will persist to find solutions without a mnemonic. They will rely on sense-making and collaboration if we create the conditions for them to do so. If math stories were a genre we introduced to students, and we normalized mathematizing stories during the literacy block, the reading and comprehension skills taught could transfer to the math block. Readers are mathers and mathers are readers!

Something else to consider is the power of storytelling. Do the students have to do all the heavy lifting of reading complex texts in order to solve math problems? Won't the reading get in the way of the mathing? No. And, no. Neither of these things have to be true because storytelling takes on different forms, and written storytelling is only one of them. Oral storytelling traditions bring tremendous value to our students. These traditions often have cultural connections and are familiar to students from diverse backgrounds. Everyone loves a good story told with emotion—and of course animated voices.

Source: istock.com/bpperry

Oral traditions have power and play a role in building community in class. Teachers can captivate students' attention and create joyful experiences with words in math class by leveraging oral storytelling. A wonderful advantage to oral storytelling is that we can insert ourselves and our students into the stories. We can also invite students to summarize or retell stories in their own words and we can leave out the quantities initially. This helps to grab the attention of the audience and puts the focus on the context and the characters, instead of focusing on which numbers to pluck out to perform random computation.

My students often believed I needed help solving a problem in real time because I prepared the story problems in advance based on events happening in our community. For example, during the morning assembly, one of the fifth-grade classes arrives late. When we return to the classroom, my wondering is about the total number of students who were present before the fifth graders arrived. Even though I wasn't sure which class would arrive last, I knew I could craft a story problem based on the number of students in the assembly. Whether it was about planning a field trip; the number of chairs needed for teachers, parents, or guests; or the number of rows the kindergarten students would take up, there were ample opportunities to establish a realistic context that my students would connect to and understand.

In preparation, I established the storyline and added in the details later. Using large sticky easel pad paper, I created an anchor chart, which is a visual tool used to support students with key concepts, vocabulary, and examples they can refer to later. On the anchor chart there were a few sentences with blanks for the quantities and possibly a question, but sometimes the students generated the questions that could be asked. When we met at the carpet at the start of the math block, I was ready to share with my captive audience, and my students could help me solve a real problem.

When sharing stories using oral traditions, we can also extend a story that students have heard before by reminding them of settings and characters they have been introduced to and understand. We can easily take familiar characters on new adventures, so students don't get bogged down getting to know them. They can enjoy the new details and jump into finding solutions to support their "friends." Students develop agency, as they are empowered to navigate complex problems and find creative solutions. Supporting students with making sense of stories before strategizing how to solve problems builds student efficacy because the context isn't interfering with selecting the methods they will employ. As they build skills and make decisions about when to use a specific math strategy, they can lean on the hints the context gives instead of hunting for key words. Imagine reading a book in the morning during your literacy block and returning to it in the afternoon during math time to solve problems with familiar characters. Mathematizing stories is a powerful way to leverage literacy for mathematical understanding and help students make connections to texts and the world.

> Mathematizing stories is a powerful way to leverage literacy for mathematical understanding and help students make connections to texts and the world.

>  **ACTIVITY TO TRY**
>
> *Math Story Time*
>
> Write a story problem on chart paper in advance and cover it up. Have the students join you on the carpet and tell the story in your own words. You can give background information, add more details, and make the story more interesting before unveiling the math task.

Visual storytelling is another option for getting students' attention and giving ALL students a chance to make sense of a context without the challenge of decoding and comprehending words. Starting a math story with an intriguing image is an example of visual storytelling. Notice and Wonder, which was formalized as a routine by Annie Fetter and the group at the Math Forum in 2007, can be used to launch the lesson, and students' ideas can enhance the story you are prepared to share. We can also use a **"zoom in, zoom out"** affect like the concept highlighted in the book *Zoom* by Istvan Banyai to create curiosity and wonder before sharing the story we have in mind.

TIP

Zoom in to show a small portion of an image and give students the opportunity to share their observations and predictions. Reveal a little more and collect their insights. Finally, show the whole picture and continue the conversation. How close were our predictions?

Source: istock.com/kali9

Math stories about the number of children at the party, the animal shape of the piñata, the number of prizes that fell out, the colors of the piñata, and more can be explored, not to mention the opportunity to discuss a cultural celebration tradition.

Mathnote
Priests, Pots, and Piñatas

One theory about the origins of the piñata comes from Aztec and Mayan civilizations of the 14th century. As a part of a religious ceremony, Aztec priests filled a container or clay pot in the shape of an animal with fruits, corn, and flowers, and decorated it with feathers. The priests used a stick to break it open during the ceremony, leaving the contents as an offering. Today, many adopt this tradition as a way to celebrate birthdays and other special occasions. How many treats does it take to fill a piñata?

Source: istock.com/Grafissimo

Whether we share written stories or visual ones, stories can support students with problem-solving if we have the right stories, and students are interested in being the hero in them. This begins with embracing the ideas that we can mathematize any story, and we have the power to create the stories our students want to hear. We also need to normalize seeing words in math class and pulling from our literacy toolboxes to investigate math stories. Storytelling traditions have been used in classrooms for ages, but mostly to teach reading and writing. It is about time we bring these storytelling traditions and literacy instruction practices into math spaces to teach mathers how to make sense of stories and the world by solving problems for characters and in their lives. Let's promote The Habits of Good Learners, so students don't turn off their sense-making superpowers when they are tackling complex texts outside of the literacy block. We want to nurture and develop literate and numerate citizens who are competent readers, writers, and mathers.

USING CONTEXTS TO SUPPORT CONCEPTUAL UNDERSTANDING

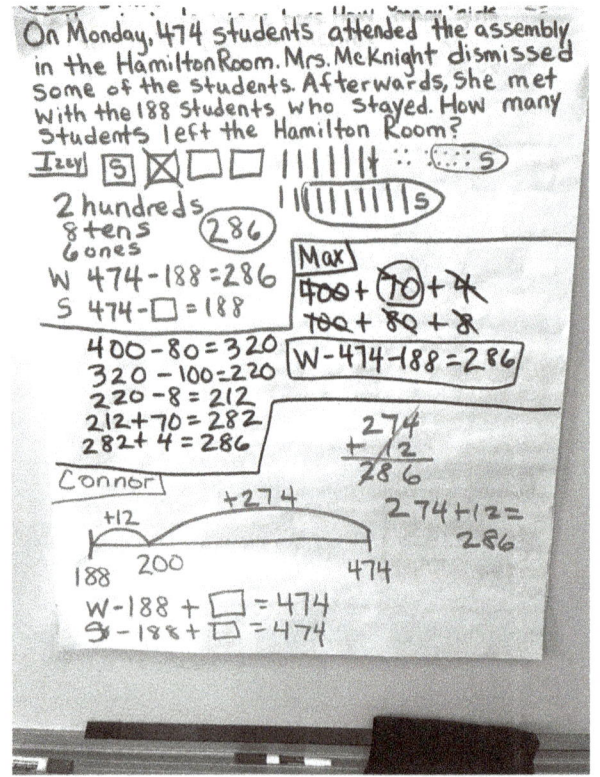

Cognitively Guided Instruction (CGI) is a framework based on more than 30 years of research and observations of children. Proponents of CGI posit that children are naturally curious, and if educators listen carefully and ask the right questions, we can learn from student thinking and anticipate the strategies that will naturally develop.

The former National Council of Teachers of Mathematics president, Mary Montgomery Lindquist, described the critical elements of CGI in the foreword of *Children's Mathematics: Cognitively Guided Instruction*. In a nutshell, she explains that CGI is focused on sense-making, is aligned to the standards, is grounded in research, respects the thinking of children, is guided and enriched through questioning, and is a quest for mathematical knowledge and experiences for the benefit of all (Carpenter et al., 2015). If you were to visit a CGI classroom, you would find students grappling with complex mathematical ideas and comparing strategies they've used to solve problems they understand because of relatable contexts. You would hear children discussing their methods and supporting their reasoning with models, visual representations, inventive algorithms, expressions, and equations. Don't we want this for all students?

At the heart of CGI is the belief that contexts play a major role in students understanding mathematics conceptually. By presenting problems embedded within stories, students strive to make sense of different types of problems and experiment with varied strategies and approaches to find solutions. By taking the time to understand the structures or types of problems, educators can use math stories to guide students toward more sophisticated strategies by managing the difficulty levels of problems. The foundational problem types for whole number addition and subtraction, which are meant to be addressed in kindergarten through second grade, can set children up for the necessary shifts from counting strategies to additive thinking and later multiplicative and proportional reasoning. A major problem that exists in the elementary grades is lack of exposure and practice with **all** types of problems.

> **TIP**
>
> Choose a picture book you've read with students or a topic of interest and write math stories for each situation type. Work with a colleague to share the load, swap stories, and solve to anticipate the strategies you will see.

Don't panic! Elementary educators need to understand problem types to effectively select or write math stories, but they need not be well-versed in 30+ years of research. It is important to trust the research that tells us that young children naturally develop strategies connected to different types of problems. A good place to start is to work with your colleagues to explore the situation types, which can be found in any CGI resource. Similar guidance can be found in the Common Core Standards or other math standards focused on approaching word problems. The beautiful thing is that identifying situation types can support teachers with task selection and differentiation and it can also help us anticipate the strategies we expect to see from students. This information is essential for purposeful planning. As teachers work through problems themselves to explore all of the possibilities, they can prepare to ask students questions that will push their thinking.

Imagine you have read *Dragons Love Tacos* by Adam Rubin during story time. Now, you would like to extend the story to practice addition word problems. Let's share stories about dragons and tacos that fit some of the addition/subtraction types. See Table 3.1 for sample story problems. While these are classified as addition situations based on the context of the story, students can use addition or subtraction to find the missing values for some of them.

Source: istock.com/tanda_V

3.1 Using a Picture Book to Practice Word Problems

Situation Types	Math Stories Based on *Dragons Love Tacos*
Join, result unknown 17 + 18 = ?	The yellow dragon ate too many tacos. She started with 17 tacos and the pink dragon gave her 18 more tacos. The yellow dragon ate them all! How many tacos did the yellow dragon eat?
Part/Part/Whole, whole unknown 78 + 63 = ?	At the dragon party, there was a boat filled with tacos. 78 tacos had cheese, lettuce, and tomatoes. 63 tacos only had lettuce and cheese. How many tacos were in the boat?
Part/Part/Whole, both parts or addends unknown *There are many possible solutions for this type of problem.*	At the dragon party, there were 25 tacos for the little dragons. Some with cheese and some without. How many of each could be on the two serving trays for the little dragons?
Join, change unknown 57 + ? = 73	Lily the dragon had 57 tacos. Sasha the dragon gave Lily some tacos, and now she has 73 tacos. How many tacos did Sasha give to Lily?
Join, start unknown ? + 34 = 83	Mike the dragon had some tacos. Gary gave him 34 tacos and now he has 83 tacos. How many tacos did Mike have at first?
Compare, quantity unknown 45 + 19 = ?	Jessie the dragon has 45 tacos. Liza has 19 more tacos than Jessie. How many tacos does Liza have? *Differentiation examples*: (45, 19) (15, 29) (145, 119)

READ. WRITE. MATH. CONNECT.

Teach students to echo the question with a blank or box for the missing value to write their final solution sentence before they solve. For example, "How many tacos did Sasha give to Lily?" Sasha gave _____ tacos to Lily. *Remind students to start with a capital letter and use correct punctuation when writing the final solution sentence.

Source: istock.com/kbeis

Depending on the grade level, the quantities can change. Even within a class, you can provide different quantities and allow students to choose. You could also assign the same problems with varied quantities to differentiate for students. Based on the images in the book, we could go with very large quantities. If we are focusing on smaller quantities, it might be necessary to build on the context since the images won't match up. For example, you could say, "What if we are hosting a party for little dragons who cannot eat as many tacos as the full-grown dragons?" Then, you start your math stories.

Many curricula include word problems, but rarely do they disperse all of the situation types equally. This results in students getting lots of practice in similar types of problems, often result unknown, with limited exposure to the more challenging change or start unknown situations. If educators quickly recognize there is a limited representation of certain types of problems, it is easy to adjust the language to rewrite or revise them. This requires practice, but the time spent will be worth it. Whether you believe reading, writing, or mathing to be your strength, you have an opportunity to grow throughout this process, especially if you partner up with a colleague or bring the challenge to your next Professional Learning Community (PLC).

When launching tasks, monitoring student progress, or sharing during the synthesis of the task, it is important to avoid absolute language like, "For this type of problem you should subtract." Or, "It will be easier if you just subtract." As a matter of fact, students do not need to learn the problem types; the teachers do. Students make sense of the story and decide the strategies they will use based on where they are in their learning trajectory. If we push or rush them to another strategy, we disrupt sense-making and run the risk of proceduralizing inventive student strategies. It is essential to honor students' thinking and strategies and allow them to adopt new, more sophisticated methods as they make connections between theirs and their classmates'. We want students to move flexibly between addition and subtraction based on what they know about the relationship of these operations. And, students need the opportunity to discover new methods for themselves, so they understand why it works and collectively make conjectures about patterns they notice with their classmates.

>  **ACTIVITY TO TRY**
> *Write Math Stories*
>
> Choose a book to read aloud, share a story your students are familiar with, or select a topic you are covering in another subject area. Write math stories for each problem type for addition and subtraction to weave in throughout the week. Don't forget to solve them in advance using different strategies you would expect to see from students.

VISUALIZING FOR MATHEMATICAL UNDERSTANDING

When students are learning to read independently or listening to a story being read aloud, they are encouraged to "make a movie in your mind" to help them visualize the characters and the setting. If children are expected to synthesize the words they've read or the details of the story they've heard, they are offered graphic organizers to help them streamline their ideas. This supports students with making sense of the story, selecting the most relevant information, and planning their writing. Whether students are writing a paragraph, an essay, or a research paper, they are taught to take the time to sift through the many details, sequence their ideas, and plan their approach for capturing the essence of the text in a way that makes sense. If we lean on these skills to solve problems in math class, students will recognize they are equipped to handle complex word problems if they treat them like math stories.

So, what might a graphic organizer look like in math class? While there are many ways to represent mathematical ideas visually, one powerful tool specifically used to make sense of story problems is the bar model. A bar model is a visual representation of a math problem using different sized rectangles to represent the quantities in the story. The bar sizes should represent the quantities proportionally to demonstrate the relationship between the values in the story. Students will be more successful solving problems if they are taught to make sense of the story **before** they start thinking about how to find the solution. For those who are not yet believers in the power of the bar model, I implore you to take the time to play around with it for yourself.

Addition and Subtraction

Part/Whole Models:
- Put together/Take apart
- Result Unknown
- Change Unknown
- Start Unknown

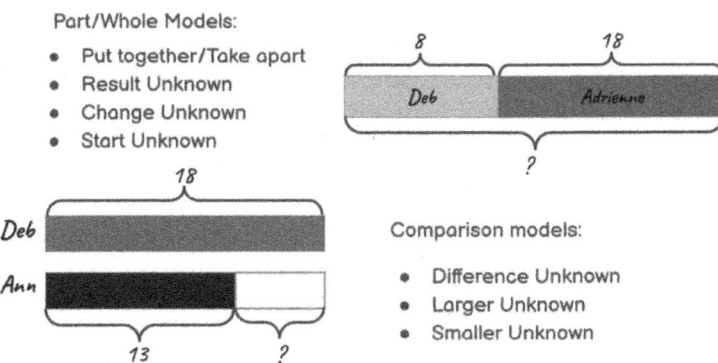

Comparison models:
- Difference Unknown
- Larger Unknown
- Smaller Unknown

When first introduced to the Singapore approach that used bar models, my colleagues and I spent time solving fourth- and fifth-grade problems with the caveat that before solving we had to draw bar models and compare our representations to one another. This was necessary and reassuring because it is challenging to teach students to use them if you are not comfortable with using them yourself or convinced of the advantage it provides. Let me emphasize that bar models are not tools for solving the problem. They are used to make sense of the story, sift through the many details, attend to the actions and number relationships, and plan the approach for finding solutions in a way that makes sense.

> *Students will be more successful solving problems if they are taught to make sense of the story **before** they start thinking about how to find the solution.*

Let's revisit our dragons and tacos in Table 3.2 and use bar models to represent each of the stories.

Source: istock.com/kbeis

CHAPTER 3 • LET'S SHARE STORIES NOT PROBLEMS!

3.2 More Math Practice Based on a Picture Book

Situation Types	Math Stories Based on *Dragons Love Tacos*
Join, result unknown 17 + 18 = ?	The yellow dragon ate too many tacos. She started with 17 tacos and the pink dragon gave her 18 more tacos. The yellow dragon ate them all! How many tacos did the yellow dragon eat? Too Many Tacos 17 18 [Yellow Start] [Pink Gave More] ? = _____ The yellow dragon ate ___ tacos in all.
Part/Part/Whole, whole unknown 78 + 63 = ?	At the dragon party, there was a boat filled with tacos. 78 tacos had cheese, lettuce, and tomatoes. 63 tacos only had lettuce and cheese. How many tacos were in the boat? Boat Load of Tacos 78 63 [Cheese, lettuce, tomatoes] [lettuce + cheese] ? = _____ There were ___ tacos in the boat.

Situation Types	Math Stories Based on *Dragons Love Tacos*
Join, change unknown 57 + ? = 73	Lily the dragon had 57 tacos. Sasha the dragon gave Lily some tacos, and now she has 73 tacos. How many tacos did Sasha give to Lily? **Sharing Tacos** 57 | ?=___ Lily Had | Sasha Gave 73 Sasha gave ___ tacos to Lily.
Join, start unknown ? + 34 = 83	Mike the dragon had some tacos. Gary gave him 34 tacos and now he has 83 tacos. How many tacos did Mike have at first? **Mike's Tacos** 83 Mike Had | Gary Gave ?=___ | 34 At first, Mike had ☐ tacos.

(Continued)

(Continued)

Situation Types	Math Stories Based on *Dragons Love Tacos*
Compare, quantity unknown $45 + 19 = ?$	Jessie the dragon has 45 tacos. Liza has 19 more tacos than Jessie. How many tacos does Liza have? *Differentiation examples*: (45, 19) (15, 29) (145, 119) Comparing Tacos Jessie Tacos | 45 | More 19 Liza | ? = Liza has ___ tacos.

In preparation for the launch of a math lesson, I have selected a context that my students are familiar with or will relate to. In this case, let's pretend we have read the story about dragons and tacos, so my students will be excited to revisit the characters they have come to love.

During my planning, I have done everything I will expect my students to do. In my math notebook, I have drawn a bar model, written a number sentence to represent the story, used a few strategies I expect to see, written expressions and equations to represent the strategies, and written a solution sentence using correct capitalization and punctuation.

When a lesson was launched with a math story, my students knew what to expect. In the case of revisiting the dragon story, I would remind students of a silly moment in the story and ask what they remembered about the dragons and their love of tacos. In my classroom, students often engaged in healthy debates about what was happening in a story and what it meant for their approach for solving. We had these discussions at the carpet before students were sent on their way to find the solution. If it was a relatively new problem type, I might draw the bar model on chart paper to create an anchor chart based on what students shared. Once they were comfortable, students could draw the bar models on their own for the launch problem and the practice problems that followed.

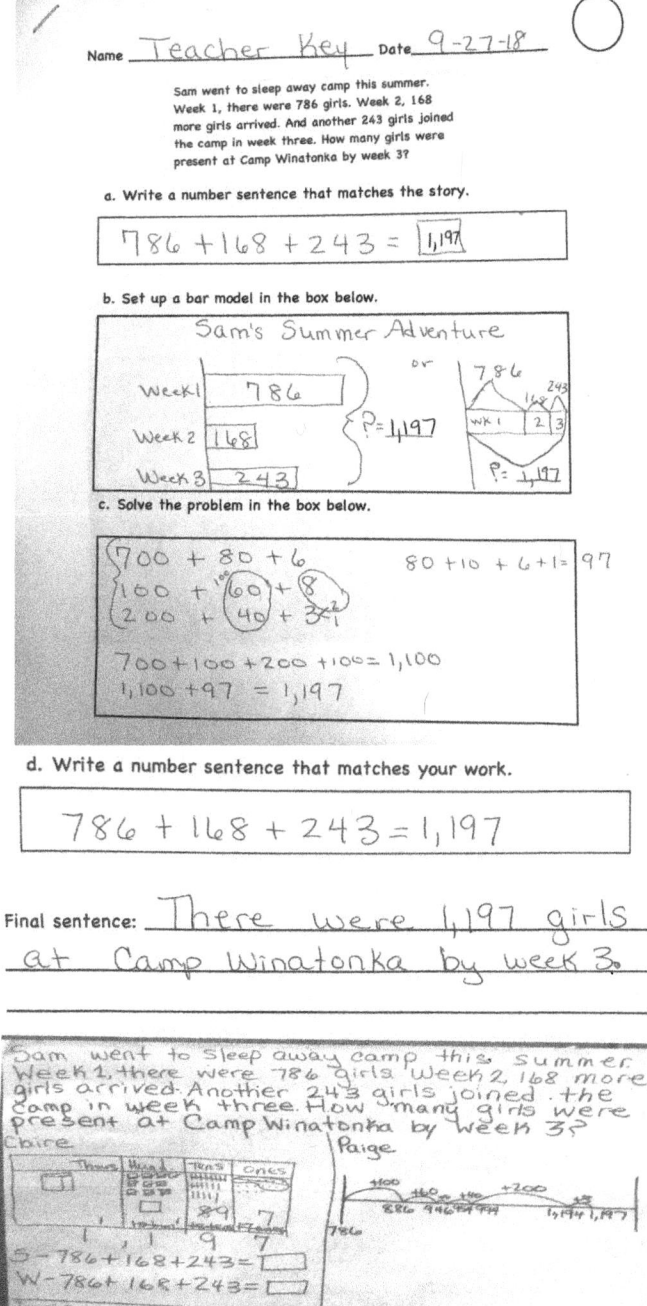

CHAPTER 3 • LET'S SHARE STORIES NOT PROBLEMS!   71

When presented with this story,

> Lily the dragon had 57 tacos. Sasha the dragon gave Lily some tacos, and now she has 73 tacos. How many tacos did Sasha give to Lily?

and asked to predict whether the solution will be greater than or less than 57, you might hear, "No, this wouldn't be a take-from because nothing is leaving. They started with some and received more. I think the answer will be greater than 57." Another student might add, "The number sentence for the story is 57 + ? = 73. Lily received more, but we don't know how much more. I think the solution will be less than 57 because we know 73 is the in all, and 57 isn't that far away from 73. Even though it's addition, I might subtract to find the missing part." In response to this, I would draw a bar model to help students consider what both students have shared and support sense-making for all students before they returned to their seats to work on the problem independently. During our debrief, we would meet at the carpet again to discuss strategies and tools that helped them find a solution.

> **TIP**
>
> To save time, instead of having students copy story problems, print them out on labels. Have students hold up a finger to receive the label and place it at the top of the workspace in their math journals or notebooks.

Another option to support sense-making comes right from the literacy toolbox, a retelling story board, a graphic organizer to help students think about the sequence of events. First Lily had 57 tacos, next Sasha gave Lily some more tacos, then Lily had 73 tacos. Some students will draw detailed pictures to retell the story, which helps them visualize the action. Of course, we cannot do this all the time, but sometimes it's a worthy detour to let students make creative storyboards. Maybe even put them on display for a gallery walk or select a few so they become an anchor chart for a specific problem type.

READ. WRITE. MATH. CONNECT.

Leverage story books across all content areas. We can teach reading lessons, math lessons, and life lessons using literature. Any story we read to children has connections to mathematics if we look for them. Preview stories to be used for anchor texts and story time with a mathematizing lens and tag the places to highlight for children. Soon, they will be mathematizing all the stories they enjoy.

When we wrapped up our time at the carpet, students received their labels with the math story and had independent work time. After a few minutes, I would offer the option to chat with a classmate and compare strategies quietly, so those who wanted to continue to work alone could. As they worked and discussed their strategies and solutions, I circulated the room "ear hustling," listening for student strategies that I knew would be beneficial to share with the group. As I made my rounds, I handed out number cards (1-4) to students who would share their approaches with the class. I anticipated a range of strategies to have students share in sequence from least sophisticated to most sophisticated to highlight connections across approaches and to nudge students to build on their go-to strategies and try new ones. This delicate push in their understanding ensured that they weren't feeling pressured to

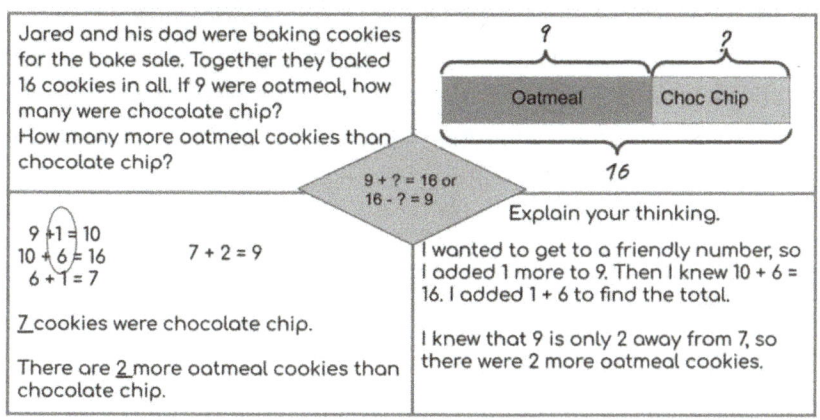

prematurely use strategies they didn't understand. Since I knew what to expect, I didn't need to hand out cards in number order. I could hand out card #4 once I saw the strategy. Students saw this as the presentation order without feeling their method was less important than another one.

During this work time, there was also an opportunity for some formative assessment, as partial conceptions were revealed, and I witnessed students helping one another to make sense of problems when they had different solutions. I often leaned in or kneeled next to a small group and asked questions to confirm their understanding and build confidence. These were questions that I had prepared in advance to push students without stripping them of their agency. Sometimes the best question to ask was simply, "Can you tell me more about that?" or "Will you explain what you did here?" Taking notes for the lesson synthesis that would help students solidify their understanding before writing their learning reflections in their notebooks was another benefit of hearing from students around the room. Even students who were not selected to share with the class that day could contribute to the closing discussion, which took place back at the carpet. We must partner with students on their learning journey and ensure that they trust the mather in themselves by allowing them to talk to one another, reason together, and develop collective agency.

> We must partner with students on their learning journey and ensure that they trust the mather in themselves by allowing them to talk to one another, reason together, and develop collective agency.

Eventually, students learned to write their own math stories. Not only was it rewarding for me to see their confidence blossom, but they were proud to see their problems show up as "homework" or on an assessment. Students wrote the problem and provided the answer key using the template I provided. The level of understanding demonstrated was off the charts because students considered how the context matched the expression, posed questions that could be answered, included a visual representation, shared at least one strategy, and explained their thinking. These stories sometimes gave us a glimpse into students' lives based on the contexts they chose. Sometimes they were complete fiction but made mathematical sense. These students proved over and over that readers read story problems, writers write story problems, and mathers make sense of problems and solve them in all different ways.

Becoming Authors of Math Stories

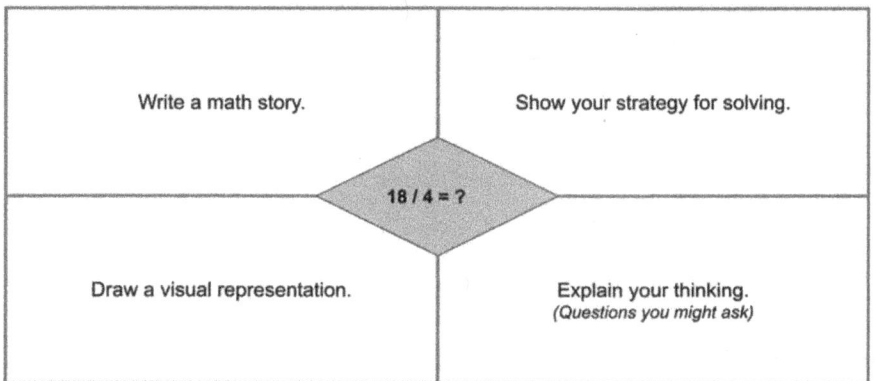

Students' voices should be amplified, as we facilitate learning from the wings. Our students deserve to be the main characters in their own stories. It is our job to give them the tools to mathematize all types of stories and nurture their curiosity so they can confidently use mathematics to make sense of the world. When we encourage discourse and debate in math class, students learn to question the ideas of others and justify their own reasoning, which are skills we expect in literacy. By combining the core principles of CGI with a visual representation for sense-making, students are set up for deeper understanding.

Keep these things in mind as you launch math time with story problems and support students with building conceptual understanding of mathematical ideas embedded within stories.

1. Use contexts for conceptual understanding.
2. Attend to and select situation types to solicit desired outcomes.
3. Plan by doing the work of the lesson and anticipating student strategies.
4. Launch math stories with routines that encourage sense-making.
5. Listen to and learn from student thinking.

> **TIP**
>
> Give students an expression or equation and have them write a story problem. Provide guidance so they include a visual representation, questions that could be answered by their story, their strategy and solution, and an explanation of their reasoning. Maybe students can even create math stories to share with a family member, neighbor, or friend and have them solve it.

6. Allow students to collaborate and compare strategies.

7. Select students intentionally to share their strategies and explain their reasoning.

8. Serve as a scribe to model organized representations.

9. Facilitate from the side and encourage student discourse.

10. Support students as they make connections between strategies and craft their own stories.

With a chapter focused on word problems, knowing that they sometimes cause stress and anxiety for students and teachers alike, it seems appropriate to end the chapter with activities to help students manage anxiety in math class and life. I discovered yoga and mindfulness for the classroom over a decade ago, and the practices I learned and shared with my students made a difference for us all, not only in math class. There are many ways to incorporate stillness, focus, and breathing in our classrooms, which equip students with self-awareness, self-monitoring, and self-regulation skills. It also brings balance to the classroom, as students learn to laugh out loud when having a good time and "Bring it in!" when it is time to settle.

ACTIVITY TO TRY
Taking Time for Mindful Moments

Students need time to transition from one activity to another and settle their minds and bodies to get ready to learn. Mindful moments are short breaks in between classes or when changing from one subject to the next. When students come back from PE and energy levels are high, that is a great time for a mindful moment. Have students sit up tall in their chairs with their feet planted on the floor, if possible. They could also sit "crisscross applesauce" with tall spines or mindful posture. Have them close their eyes or lower their gaze so they aren't distracted by others. It is important to reassure them that you will keep your eyes open to watch over them and make sure they are safe. Set a timer—a sand timer or timer with a quiet alert works best—and have students take a minute to just breathe. It is helpful to teach them a breath like Calming Breath (see next page) or encourage them to breathe in and out slowly and evenly, focusing on their inhales and exhales. You can even have them count their breaths silently in their minds to maintain focus. I sometimes had students count as high as they could in their minds and asked how far they got when we were done. Other times, I encouraged them to count backward from a number and asked the same question. Initially, 30 seconds might be enough but over time, they get better at this practice and 1 or 2 minutes offer a much-needed break.

Mathnote
Mathing While Breathing

Calming Breath is a mathematical breath. This breath calms the sympathetic nervous system and helps us to slow down. You can inhale for a count of two and exhale for a count of four, in and out of your nostrils. If students feel more comfortable exhaling through their mouths, it's okay. They can purse their lips like they are blowing through a straw. By taking twice as long to release your breath, you can regain control if you are feeling out of sorts. The more you do it, you may find you can inhale for longer and exhale for twice as long. If you inhale for three counts, you would exhale for six counts. Just think about doubling and halving to remember the calming breath formula. Yogis math!

ACTIVITY TO TRY
Calming Glitter Jars

Calming Glitter Jars are a fantastic tool for students to make or have available for focus and settling their minds and bodies. As they shake the jars and watch the glitter settle, they are reminded of their ability to sit still and settle their thoughts. The flurry of colors mingling and floating through the water is the perfect focal point when trying to tune out distractions.

(Continued)

CHAPTER 3 • LET'S SHARE STORIES NOT PROBLEMS!

(Continued)

Here's how you can make your own:

1. Recycle a glass (or dishwasher safe plastic) jar. Make sure it is clean and dry. A smooth clear jar with a screw-on lid works best.

2. Pour a little bit of water (room temperature) into the jar, so the glue doesn't stick to the bottom of the jar.

*****For steps 3-5 an adult MUST be involved.*****

3. Have an adult boil or heat up water to fill the jar. I use an electric kettle. Pour the hot water into the jar, filling about $\frac{1}{4}$ of the jar.

4. Squeeze two or three tubes of glitter glue into the hot water.

5. Cover the jar and shake the bottle gently in a circular motion until the glitter glue separates and the glitter starts to settle.

6. Open the jar and add enough room temperature water to finish filling the jar.

7. Screw the lid back tightly onto the jar and shake.

8. You can decorate the jar with paint pens and use self-adhesive craft sheets to decorate the top of the jar for added pizzazz!

9. Watch the glitter settle and enjoy!

 **ACTIVITY TO TRY**
Incorporating Mindful Movement

Young students do not always do a good job of controlling their body Movements. In PE, they are learning how to build their muscles and how to make their limbs do what they actually intend for them to do. A terrific way to support the development of gross motor skills in the classroom is to incorporate mindful Movement and brain breaks. Getting the wiggles out is sometimes necessary before launching into the next lesson of the day. There are resources available to find creative brain breaks for students, like GoNoodle.com, that you can play in your classroom.

Here are a few ideas from my classroom:

- **Slow Motion Up and Down:** Using a sand timer or another visible time keeping instrument, set the time for one minute and task students with using the entire time to get up from a seated position or to sit down from a standing position. The goal is to stay in motion without moving too fast or too slow and land or stand perfectly in sync with the timer.

- **Take a Mindful Walk:** Students enjoy taking a mindful walk as a break or on our way to lunch or the playground. We leave the classroom with our mindful ears on, listening for sounds we often ignore or are too busy to take in. We look around with fresh eyes, noticing our surroundings. Students walk at a slow, steady pace in silence with their senses on high alert. Later we share about the things we noticed during our mindful walk. A student once shared with excitement that he heard a bug fly by and he was sure he heard the sound of its wings.

GoNoodle
https://gonoodle.com

WHERE'S THE MATH IN THAT?
Yogis Are Mathers! (My Story)

According to Luttenberger and colleagues (2018), math anxiety is one of the most prominent forms of anxiety connected to a knowledge domain. These researchers categorize math anxiety as a widespread mental health issue that affects people of all ages all around the world. Environmental factors

(Continued)

(Continued)

such as stereotypes, attitudes of adults, math ability, negative experiences, and personal characteristics all serve as antecedents that facilitate the development of math anxiety. Maloney and Beilock (2012) define math anxiety as "an adverse emotional reaction to math or the prospect of doing math. Despite normal performance in most thinking and reasoning tasks, people with math anxiety perform poorly when numerical information is involved." Beliefs and mindsets about who is good at math, negative consequences for poor performance with math tasks, and learning math concepts through procedures and rote memorization without understanding can exacerbate math anxiety.

When I noticed young children experiencing math anxiety, which sometimes paralyzed them or caused tears at the mere mention of math, I knew something had to be done. I made the decision to attend a training that focused on bringing mindfulness to the classroom. Their approach incorporates five key elements in each session:

- Connect
- Breathe
- Move
- Focus
- Relax

Source: Michael Moye

Through these elements, and within a framework of respect for all students, compassionate communication, and joyful exploration, students are taught fundamental life skills that are immediately available for daily use. While the training was designed to provide teachers with tools and strategies for developing students' self-regulation and focus in general, I knew it would be helpful with alleviating math anxiety. This was the beginning of my yogi journey.

After finishing about 100 hours of training with Little Flower Yoga (n.d.) to support students in the classroom, I decided to complete a year-long 200-hour yoga certification through Yoga Haven. Later, I continued my studies with about 90 more hours of training because of the positive outcomes my students were experiencing. I wanted to understand how yoga could provide an outlet for students experiencing anxiety and I wanted to be able to answer students' questions about how contemplative practices help our brains. The lessons I learned about self-awareness, balance, and controlling emotions with breathing and Movement changed everything. Students gained confidence and approached uncomfortable situations feeling empowered and in control. This was a benefit in math class, but it spilled over into every area of their lives and mine.

I adopted mindfulness and yoga as practices of my own. A 20-minute mindfulness practice at the start of my day prepared me to be the anchor in the classroom. Students entered the room each day to find a place for centering, shaking off whatever happened the night before or on the way to school, and an environment primed for optimal learning. When they crossed the threshold, sometimes I witnessed a sigh of relief as they unpacked backpacks. Some students retrieved calming glitter jars from the shelf, or pulled out a book to read, while others simply smiled and said good morning. Finding balance and choosing peace daily was the gift I gave to my students, my family, and my community. I wanted to share this way of being with as many people as I could, so when other teachers noticed the difference and inquired, I taught them.

We hosted grade-level workshops and family events. I offered literature-themed yoga classes and explored shapes through yoga poses. Students learned how to use their breath to regulate their emotions and how to settle themselves before a math test. They learned to be better listeners and to attend to their feelings without being reactive, and they taught the strategies they were learning to their family members.

Yoga and mindfulness changed my life. It became a part of who I am. I taught yoga classes in studios, after-school programs, summer programs, and community centers. One summer, students called me Peace Lady when they couldn't remember my name. Students visited The Peace Room for literacy, life lessons, and yoga. It seems strange to say, but I became a yogi because of math! Yogis are mathers because, much like everything, math is an important part of a yoga practice. We can find angles all around us, including in yoga poses! Even if you have no intention of becoming a certified yoga instructor, there are practices that you can bring into your classroom to help students manage anxiety in math class and life.

KNOW YOUR ANGLES

Deborah Peart, our featured mathematician, knows that yoga is an excellent exercise for your body and can be very relaxing. *Asanas* (yoga poses) require our bodies to create different angles. Trace over the angles of each pose and write whether it's straight, right, acute or obtuse. If you have a protractor, measure each angle to the nearest degree. Measure as many angles as you can!

Safety Note: Do not try these poses without adult supervision/a trained professional!

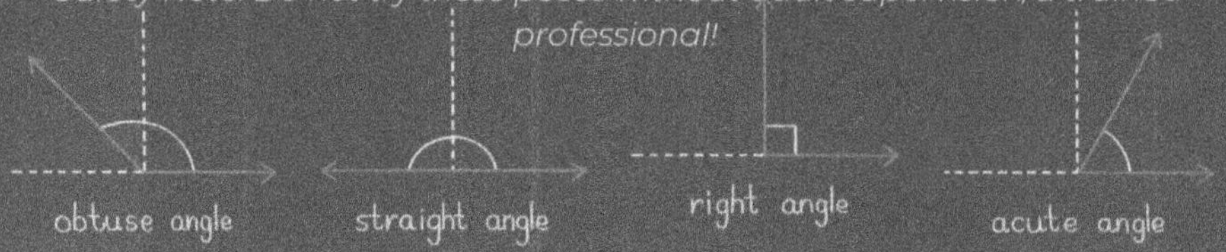

obtuse angle · straight angle · right angle · acute angle

Source: Michael Moye

TIME TO REFLECT AND TAKE ACTION 3

Sometimes we believe something because it is what we have always known to be true, but once it is challenged and our eyes are opened, we can't go back. If we believe that we are not "math people" or that we are only good at teaching literacy concepts and strategies, what does that mean for our students? It is time to move past acceptance and recognize that we have the power to change our mindsets and beliefs about what it means to be good at math. I hope this chapter helped you consider the strengths you have for teaching literacy strategies that are transferable. This means you have everything you need to teach mathematics with the same confidence, but it will take some time, practice, and a few deep breaths.

Word problems have the reputation of being scary. They are the "optional" tasks at the end of the lesson that we don't mind skipping. If we shift our perspectives, we can celebrate stories and help students see the joy that can be found in all kinds of stories, even in math class. What are the math stories waiting to be written by YOU? It is time to embrace your literacy strengths and stretch yourself in mathematics . . . and maybe incorporate a little yoga stretch every now and then too!

1. What strength with teaching literacy will you leverage when teaching math?
2. Which literacy routines can you implement to support math learning?
3. How will you support students with the transfer of literacy skills to making sense of math stories?
4. What is a favorite story that you will approach as a mathematizing opportunity?
5. How can you support families with mathematizing stories at home?

MATHFIRMATION

Students need to recognize that they might enjoy reading more than mathing, but it doesn't mean they are not mathers. Have students stand in a circle or in a line and take a step forward for the statements that are true for them. Have them repeat the statements confidently. Use a few, mix up the order, add to the list, and help students recognize their ability to work hard at learning necessary skills whether or not they love them.

(Continued)

(Continued)

Statement #1: I can read, so I am a reader.

Statement #2: I LOVE reading.

Statement #3: Reading is not my favorite, but I am not afraid to do it.

Statement #4: Reading is sometimes hard, but I can do hard things.

Statement #5: I need to be a good reader to be successful, so I will practice every day.

Statement #1: I can write, so I am a writer.

Statement #2: I LOVE writing.

Statement #3: Writing is not my favorite, but I am not afraid to do it.

Statement #4: Writing is sometimes hard, but I can do hard things.

Statement #5: I need to every day.

Statement #1: I can ma

Statement #2: I LOVE n

Statement #3: Mathing

Statement #4: Mathing

Statement #5: I need to every day.

PART 2
MATHERS GONNA MATH!
Taking Math Outside of the Math Block

Source: istock.com/PeopleImages

MATHEMATIZING ACROSS CONTENT AREAS

Source: istock.com/Kobus Louw

I recently had the opportunity to observe sessions of high-impact math tutoring with first and second graders who were pulled out for a 30-minute, one-on-one session. There were about 10 children working with volunteers, practicing foundational skills to support the concepts they were learning in class and to build their number sense. Students reasoned their way through tasks, sharing their developing mathematical ideas, and played games with their tutors to reinforce the skills. Across the room, there was laughing and cheering with every victory finding combinations to make 10. As the time wrapped up, one student exclaimed, "I like this math. It's fun! It's not like regular math." Sadly, this is not a unique reaction. There is a pervasive belief that mathematics by definition is not supposed to be fun that starts in elementary school and continues well into the upper grades and adulthood. There is a disconnect between how mathematics is experienced in classrooms and what mathematics is actually meant to be.

> There is a pervasive belief that mathematics by definition is not supposed to be fun that starts in elementary school and continues well into the upper grades and adulthood.

In most elementary schools, classes are self-contained with a teacher who is responsible for teaching the core academic skills, reading, writing, and mathing. They are often expected to also teach science, social studies, and some form of character education. Unlike in the upper grades where teachers focus on a specific content area like mathematics or English, elementary educators are generalists and required to do an excellent job across all content. You all are superheroes! But this expectation to know it all and do it all also means that elementary educators are faced with making regular decisions about what to prioritize for student learning. Especially when time is short, these in-the-moment decisions are often guided by our strengths and preferences. In other words, those of us who love reading rarely cut reading time. Conversely, those of us who don't love math are often willing to adjust math time or move it to the end of the day when students are not at their best. What if we turned this around by considering cross-curricular connections and how we can weave mathematics throughout the school day, every day?

Let's consider how cross-curricular connections happen with reading and writing. Whether teaching science or social studies, reading and writing will be a part of the equation. Students transfer skills learned in the literacy block to other content areas seamlessly. No one questions why they are still talking about reading or writing during science or social studies. Why is that? The simple answer is reading and writing are NOT optional, so no one questions when it is necessary in every course and all throughout our lives. As a matter of fact, they expect it, which is why the quest to be literate citizens continues well into adulthood.

READ. WRITE. MATH. CONNECT.

Students expect to read and write in all subject areas because they have been conditioned that reading and writing are necessary to make sense of the work they are doing. If we emphasize the role mathematics plays in other content areas, students will begin to expect to engage with mathematics and to use mathematics to make sense of other subjects and the world around them.

> **TIP**
>
> Let's normalize special times for math play and exploration! Find time in your schedule a few times a year to focus on math through games and other fun activities with less structure imposed. Maybe create a special math challenge for the week or just give students time and space to enjoy math without restrictions or grades attached.

When we look to the standards, the reading and writing standards incorporate nonfiction reading and writing and give examples of how to assess reading and writing skills that are transferable across content areas. In classes outside of the literacy block, we read books and articles, write out our hypotheses, use graphic organizers to prepare write ups of our findings, and outline our persuasive writing essays to defend our positions. Students are even encouraged to read and write in their free time, positioning reading and writing as skills that are meant to support leisure activities and something you might choose to do even when not being graded. Why don't we see these explicit connections being made with mathematics? When was the last time we offered students "Drop Everything and Math" days? Mathematics needs to be positioned as both necessary and enjoyable to shift our practices.

To support students with building their background knowledge and improving reading comprehension, we intentionally seek out students' funds of knowledge and connect familiar ideas across content areas. We work to offer students mirrors, windows, and sliding glass doors, as defined by Bishop (1990). She emphasizes the need for students to see themselves and their worlds reflected in the books they read, while also having opportunities to peek into other worlds unlike their own, and find ways to make connections between them. While her focus was on literature, this idea rings true in mathematics as well. Students need to see themselves represented in the contexts of problems, but they also need to see examples of people who look like them or who come from similar backgrounds using mathematics across their lives. This could mean having a wall of mathematicians who are racially and ethnically diverse and highlighting historically marginalized people who have made contributions to the field of mathematics. This could also mean inviting in a parent who is a local carpenter to explain how math is used in construction work and carpentry. From the ordinary to extraordinary, representation matters.

Source: istock.com/Adene Sanchez

It seems ridiculous to even consider how we would separate reading and writing from other content areas, and yet mathematics is viewed as a subject relegated to the math block. If we redefine what it means to be literate to include using mathematics to make sense of the world, this could change.

Let's think about how we can change the status quo of mathematics being taught in a silo as a set of procedures and skills and move mathematics outside of the math block. Let's consider how we move away from the belief that mathematics is a set of disjointed ideas that are meant to be memorized to pass a test. Let's consider practical ways that we can bring mathematics to life for our students and drive home the point that math is NOT optional. Let's position mathematics as a powerful tool to help us solve problems and a beautiful coherent story that is meant to be enjoyed.

> Students need to see themselves represented in the contexts of problems, but they also need to see examples of people who look like them or who come from similar backgrounds using mathematics across their lives.

In recent years, STEM programs and initiatives have popped up everywhere with the mission of making STEM careers more accessible for all students. There are special clubs, classes, and camps focused on preparing children for a path that historically has been reserved for a select few. Many of these programs have cool Science projects, leverage Technology, and introduce students to the world of Engineering. While I celebrate these new avenues, I find myself wondering if the field of STEM would be better served by supporting students with developing positive math identities in the early grades. Oftentimes, students get excited about hopes of a future in STEM without recognizing the critical role success

> Let's position mathematics as a powerful tool to help us solve problems and a beautiful coherent story that is meant to be enjoyed.

in mathematics will play in their futures. Unfortunately, because of mathematics' reputation as a buzz kill, marketing for STEM programs often includes highlights of the "S," the "T," and the "E," while the "M" is left out of the promo reel.

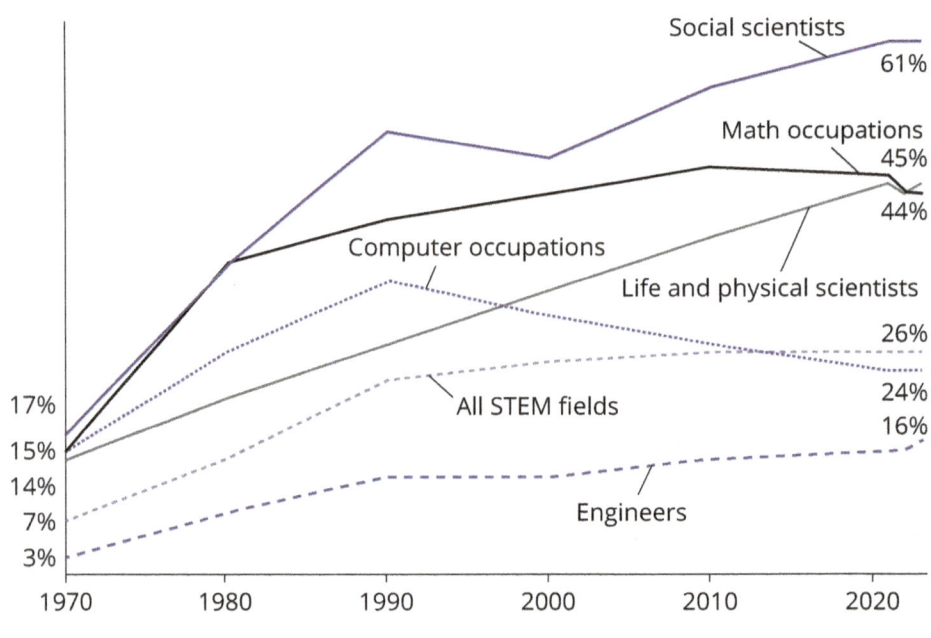

Percentage of science, technology, engineering, and math (STEM) workers who are women

Source: US Department of Labor, 2025

The landscape of STEM careers is changing at a slow pace because some of the historical biases about who "deserves" to be in these spaces and the gatekeeping to keep out any deemed "unworthy" still exist. According to the US Department of Labor Women's Bureau (2025), only about 26 percent of workers in STEM fields were women in 2020. That number increased to about 31 percent in 2023, which leaves a lot of room for improvement. According to the US National Science Foundation (2023), Black, Hispanic, American Indian, and Alaska Native people collectively made up 31 percent of the US population, but only 24 percent of the STEM workforce in 2021. Additionally, these groups typically have positions that require certifications and technical skills rather than associate's degrees or higher education and have lower salaries and higher unemployment rates than white or Asian STEM workers.

This landscape accurately reflects the disparities in high-level math courses in middle and high school, which has its roots in elementary classrooms where students begin to "opt out" of mathematics.

As students begin to accept the mistaken notion that they will never be "math people" or that mathematics success is determined by the imaginary "math gene," they lose hope and give minimal effort. Some will even develop math anxiety, defined by Maloney and Beilock (2012) as "an adverse emotional reaction to doing math or the prospect of doing math." Bates and colleagues (2011), tell us that "Mathematics anxiety stems from a lack of confidence or low self-esteem in mathematics, as well as a fear of failure or past negative experiences with mathematics." We can help students recognize that sometimes learning math concepts will be difficult, BUT with hard work and perseverance we can ALL learn mathematics and be successful.

My theory is that if we provide students with opportunities to struggle productively in a supportive environment, cultivate a culture of curiosity, and create engaging math experiences for students, they can develop positive math identities in elementary school. Over time, more students will embrace **Mather** as a part of their identities and make decisions about career paths from a place of confidence rather than a fear of "too many" math courses. Just as we never give up on students who are challenged by learning to read, we cannot give up on students who are challenged when learning new math skills and concepts. They won't all choose to pursue STEM careers, but we can ensure that they won't miss out on pursuing their dreams because of mathematics. Our goal as elementary educators is to provide ALL students with options by preparing them to view math challenges as opportunities for growth, not doors slamming on their dreams.

> *Our goal as elementary educators is to provide ALL students with options by preparing them to view math challenges as opportunities for growth, not doors slamming on their dreams.*

Source: istock.com/Marilyn Nieves

Before there was buzz around STEM, we often heard that science and math go together. There was a belief that you cannot do anything in the field of science if you aren't a whiz at mathematics, so you had to choose career paths wisely.

> "If you want to be a doctor, you will have to take lots of math courses."
>
> "If you're taking chemistry, you are going to do lots of math."
>
> "Wait until you get into your physics courses. There is so much math!"

While this may be true, it is not very encouraging to a young student who is feeling the pressure of the famous question asked far too soon, "What do you want to be when you grow up?" Whether you thought you could be a singer, dancer, astronaut, veterinarian, or a professional basketball player who occasionally performs life-saving surgeries, you didn't really have enough information to make a sound decision that would last long term. Don't get me wrong: I do believe we should have conversations in elementary school about the types of careers we can aspire to have. More importantly, I would love for us to share information about career paths without using math as a deterrent. For example, saying, "If you want to be a doctor, you will have to take lots of math courses," could be the comment that plays on repeat in the mind of the student who is struggling to understand why fractions are numbers. The problem with these broad sweeping statements is that it limits the idea of what mathematics is and can be. There will be lots of math in every career choice, so attaching it to certain paths is unnecessary and misleading.

When students say, "I'm bad at math," what responses do we offer? Too often we launch into some version of why math isn't for everyone or it's okay to not be good at it. Even though this response is often well-intentioned, we need to support students with understanding that mathematics is not singular, and emphasize the fact that it is highly unlikely that you are just "bad at math." The first thing we should ask is, "What part?" or "Which skill?" making it clear that failing at one thing in math class is not equated to failing at mathematics. Another very important thing we can do for students is show them all the math they are already doing all the time. One good place to start is in science class.

In elementary science classrooms, there is no shortage of curiosity and wonder. By calling out the math that is happening in science class, students begin to see mathematics as a tool for solving problems in science and the natural world. With the introduction of the Next Generation Science Standards (NGSS), the team of educators and experts in the fields of science and engineering were careful to consider the math concepts that align with scientific principles so the mathematics understanding would support science investigations. And much like the expectations in mathematics, the science framework focuses on understanding content conceptually and applying this understanding to engage in

scientific inquiry instead of memorizing facts that are not grounded in a meaningful context. Let's take a look at interpreting data as one topic that we can leverage for cross-curricular connections to science.

READ. WRITE. MATH. CONNECT.

We can highlight the "M" in STEM by connecting curiosity and exploration in science to mathematics. We must help students recognize how mathematics can illuminate hidden answers to our questions and support sense-making in science class. As we write up our findings and justify our theories, we can consider and connect the ways in which mathematics validates our hypotheses and discoveries.

In many curricula, the unit on data interpretation is one of the last and one that is often skipped when the time crunch of wrapping up the year sets in. To avoid this and provide more opportunities for students to engage with data, this is one of the units I introduced early and continued to build on all year. While I didn't have specific guidance from the curriculum,

I made informed decisions based on the resources available to me at the time. When I was working as the lead author for second grade at a curriculum company, we had many discussions about how powerful it could be to start the year with the data unit. We recognized that across multiple grades, this unit could be used to assess computation skills from prior grades, build community through surveys and data collection, and introduce visual representations that could be used to represent mathematical ideas. The NGSS also call for elementary students to be able to analyze and interpret data, ask questions and define problems, plan and carry out investigations, use mathematics and computational thinking, develop and use models, and construct explanations and design solutions. Alignment between science and mathematics is meant to be, but it requires us to be intentional about sharing the connections with students.

Surveys are "a sampling or partial collection of facts, figures, or opinions taken and used to approximate or indicate what a complete collection and analysis might reveal," according to Dictionary.com (n.d.). In elementary classrooms, they can be leveraged as a tool for information gathering and getting to know the members of the class community and the extended community, including other students and faculty in the school and families. When introduced early in the school year, surveys can be used regularly to poll opinions, find similarities and differences, and practice data interpretation skills like defining problems, asking the right questions, collecting data, analyzing and interpreting data, using computational thinking to answer questions and find solutions, and creating representations to share their findings with others. Sound familiar? These criteria are found in math and science standards, and when students understand and develop these skills, it supports them in future math and science classes.

Source: istock.com/joey333

Starting with our youngest learners, students are capable of having conversations about their opinions, like favorites or least favorites, and finding commonalities among their peers. As they move through the grades and learn more about how to organize and display data using charts, tables, and graphs, their organization of the data collected and representations will evolve. For example, kindergarten students often talk about the calendar, attendance, the weather, and other topics of interest during some form of a morning meeting. This is a great time to have weekly data chats. Students can manage asking family members and friends simple questions and collecting responses with the support of clear directions and collection tools. Then their data can be displayed in an organized way to guide conversations about what we can learn from our findings. Through each data representation, students collectively tell a beautiful story that includes the preferences, opinions, and experiences of all the members of the community. If done all year long, the story the data tell us will be one of growth, diversity, sameness, and belonging, one where all voices matter.

> **TIP**
>
> Make it a regular routine to ask students about their preferences and opinions and create a representation of this data in an area of the classroom where students can visit and check the results of the weekly survey data. Take time to ask questions about which category had the most or least. Practice finding the difference between categories, asking how many more or less questions, and practice basic computation skills by having students add up total amounts of categories or compute how many more a certain category needs to reach a goal.

WOULD YOU RATHER . . . ?

Students enjoy surveying family members, friends, and anyone who will listen. In my classes, we often took polls about foods, celebrations, and even bed times. Another simple activity we engaged in regularly was a Would You Rather? routine (see next page), meant for quick data check-ins. Students shared their choices, and I displayed them by making a chart or graph on the whiteboard or an anchor chart, or with sticky notes added to my graph that I prepared in advance. With this routine, it is important to keep it to two options, but allow neither as a response. It is also good to compare two things that are similar in form or function. When the choices are similar, students often grapple with their responses because they likely enjoy both. This is a wonderful chance to teach about getting only one vote. When asked, "Would you rather have pancakes or waffles?" Students protest because they love both, but both is not an option. My reply, "You must pick only one." When all students have made their selection, we have a conversation about what this tells us about our class community. Students who chose neither will often explain why they don't like the options. Maybe this serves as a chance to learn that a classmate eats other breakfast foods because of an allergy or because of their culture. This can be a window into the lives of their friends.

Source: istock.com/margouillatphotos

 **ACTIVITY TO TRY**

Would You Rather?

Try out these Would You Rather? prompts to see how your students enjoy the routine.

- Would you rather have a cat or a dog as a pet?
- Would you rather have a cookie or a cupcake?
- Would you rather have an apple or an orange?
- Would you rather walk or ride a bike?
- Would you rather be too hot or too cold?
- Would you rather read a story or write a story?

Students begin to realize that not everyone loves what they love, AND that it is okay. My students tried with great passion to convince a classmate that pizza is awesome, but she stood firm with her decision that she doesn't like it because she doesn't like sauces and cheese is not her favorite either. While they were shocked, they respected her choice. When we celebrated a class victory later in the year, and the prize offered was a pizza party, they petitioned the principal to offer another option because it wouldn't be fair to their classmate. Our class party included pizza, subs, and salads because of their advocacy.

Source: cookie slices image by istock.com/Pinkybird; cookie halves istock.com/Zakharova_Natalia

This routine is also a great way to launch into new topics in mathematics and other content areas. While we were learning about comparing fractions, the Would You Rather? that day was, "Would you rather have $\frac{1}{2}$ or $\frac{1}{4}$ of your favorite treat?" It was no surprise they unanimously voted that $\frac{1}{2}$ would be better. However, when I explained how much I loved chocolate chip cookies and I would prefer $\frac{1}{4}$ of one, they were genuinely perplexed. They wondered whether I was making my choice because I wanted to be healthy. I responded that I wanted the most cookie I could have, so I wanted $\frac{1}{4}$. Of course, they debated that $\frac{1}{2}$ is more than $\frac{1}{4}$ and tried to convince me to change my choice. I asked, "Is $\frac{1}{2}$ **always** more than $\frac{1}{4}$?" Some students did not sway, while others paused with a questioning look.

When I revealed the images of a tiny chocolate chip cookie cut in half and a giant chocolate chip cookie cake cut into fourths, and began to dance with excitement, they did protest that it wasn't fair that I had not mentioned the size difference. This was my plan all along. To compare fractions, we must consider the whole. At the end of the unit, when asked, "Would you rather have $\frac{1}{2}$ of a cupcake or $\frac{1}{4}$ of a cupcake?" the first thing students asked was whether they are the same size. When I revealed the images of a regular small cupcake and a large cupcake cake, they rejoiced because they had not been "tricked." Lesson learned!

COLORFUL CANDIES OR DANGEROUS MOON ROCKS?

If we want students to be curious and make creative connections, we need to model curiosity and creativity. So, when it was time to dig into the data unit and explore tables, bar graphs, and line plots with my third graders, I had the perfect plan. Students were tasked with sorting and organizing data in a way that made sense to them, and using tally marks in a table with the correct labels to share their findings. They had to graph the data using a line plot and a bar graph. Next, they had to summarize their findings in a paragraph and include the questions that could be answered by their data. Last, they made a creative display to share their data, including questions that others could use their data to answer. At this point in the unit, much of this was practice, so of course I found a way to challenge them all.

Source: istock.com/SasaJo

The "data" students were given was one full-sized bag of Skittles. Some students had regular, some had sour, others had tropical or wild berry. (This matters!) Because I didn't want students to just count and organize candy, there was a twist. Each student was tasked with creating their own context through creative writing. They were "in the lab" trying to determine what was going on with their data, and the paragraph summarizing their findings had to be within the context they had defined. I went around the room as they worked and interviewed each "scientist" addressing them as Dr. and their name. One of my favorite interactions was with a student who, when asked what he was doing, responded, "I am trying to determine if these brightly colored rocks are dangerous and what each color represents." When I asked where they were found, he simply stated, "I'm not sure. I just received the data."

This project spanned two days. On day one, students worked independently, but on the second day, they were given a partner and instructed to go through the whole process again. The key was that partners had different data, types of Skittles, and were tasked with merging their data and creating a joint representation. The interesting *noticings* and *wonderings* that came up were fantastic. Students contemplated how to combine data now that some colors were the same but some were different. Some students merged data by dumping out all of their Skittles, on their separate plates of course, and counted. Other partnerships confidently shared that they simply added their totals together for similar categories and created new categories for the differing ones. Some students decided to sort in a new way altogether, flavors instead of colors or whether the "S" on each piece of candy was whole, partial, or missing. Their joint displays included their individual representations and the new merged data representation.

We finished our two-day exploration with a gallery walk where students answered questions that were included on the poster, as required, using the graphs and mathematical computations. And of course, everyone put their Skittles in the baggies that were provided to take home. It was so rewarding. Even though one student chanted the whole time he was counting, "Sorting the data, not eating the data," others used rulers with precision to make sure their graphs were proportional and accurately displayed their findings. The creative stories were hilarious, including magic beans, glowing orbs, moon rocks, and more.

In this two-day task, students engaged in reading, writing, science, and mathematics, and honed skills needed across them all. Students were focused and engaged and applied their learning to create a spectacle for all to see and learn from. Even though I know that technically line plots are meant to be used for measurement data, and this context was not real, the practice making sure the Xs were all the same size, the skilled use of graph paper, and the practice asking and answering questions, were all valuable lessons. Later in the year, when we continued to interpret data, these third graders were competent and confident in their abilities and continued to sharpen their skills. They even had an opportunity to use these skills to lobby for justice in the cafeteria. (You will read about that soon.)

Mathnote
Mathing Around the Globe

Dr. Gladys West is a retired mathematician who worked for the US Naval Proving Ground in 1956. There, she solved complex mathematics equations by hand and eventually programmed computers to solve them for her. Dr. West designed a program, the Naval Ordinance Research Calculator, to track the Movements of Pluto in relation to Neptune. Later she managed the project that led to the discovery of using satellites to observe oceanographic data. Because of her work on GEOSAT, which is a satellite that was programmed to capture the shape of the Earth and create models of the Earth's surface, GPS systems can accurately calculate any location on Earth.

Slow Reveal Graphs: Watching the Story Unfold

Slow Reveal Graphs curated by Jenna Laib

https://slowrevealgraphs.com/

Now, you might be thinking, "How do we get students interested and keep them interested in data all year?" Well, I'm glad you asked. In addition to having them collect data and share about their families and friends and using the Would You Rather? routine, Slow Reveal Graphs are another wonderful instructional routine to pique the interest of students while helping them develop a keen sense about what type of information is needed to draw conclusions and interpret data. It is also a great way to have students attend to the details in different types of data representations with hopes of them learning how essential the structures and labels are for sharing information with visual representations. It is also an excellent way to make cross-curricular connections by choosing graphs that align to the topics you are discussing and learning about in other subject areas.

Slow Reveal Graphs are often used in social studies and science classes to help students make sense of data. Jenna Laib, a math specialist in Massachusetts, created a site to demonstrate how these graphs could be used as an instructional routine in math, science, and social studies. Talk about cross-curricular connections. She defines a Slow Reveal Graph as, "an instructional routine that promotes sensemaking about data. This highly engaging routine uses scaffolded visuals and discourse to help students (in K-12 and beyond) make sense of data. As more and more of the graph is revealed, students refine their interpretation and construct meaning, often in surprising ways. This routine increases access for students without sacrificing rigor or engagement" (Laib, 2024).

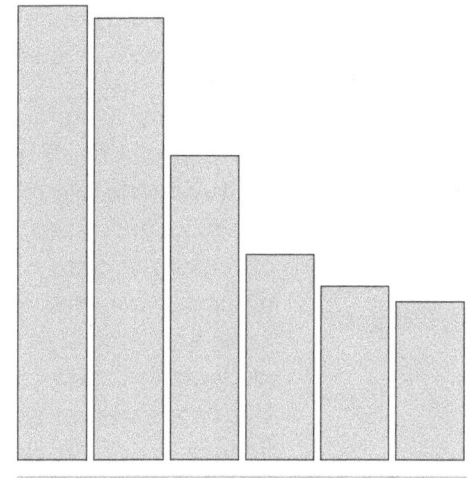

Source: istock.com/Rowan Jordan

As students notice and wonder about a bar graph missing labels, numbers, a key, and title, they consider the shape the collection of rectangles makes. Then they question what it could mean that one bar is twice the height of another. After the next slide, they speculate about what the numbers below the X-axis represent. With each new slide, more information is revealed and discussions continue, as students talk about how the new information supports their predictions or causes them to revise their thinking. This routine is another terrific method to engage students with data in a way that is interesting and helps them become comfortable with data interpretation. Data science is a hot topic now, with some advocating for it to be more present in K-12 education. Dr. Jo Boaler, a professor of mathematics education at Stanford Graduate School of Education, has been on the forefront of the data science revolution, and played a major role in the campaign to update California's math standards. Dr. Boaler and her team at YouCubed.org have curated resources to support educators with strategies for incorporating data science and preparing students to navigate the 21st century with confidence. Since we will inevitably need to interact with data and make sense of charts, graphs, and tables to understand what is going on in the world, it is

YouCubed .org Data Science

https://qrs.ly/8ugnmtl

critical to introduce the skills needed to engage with data at a young age (Boaler & Williams, 2026).

Slow Reveal Graphs open the data discussion to all students without the intimidation of having the wrong answer because it is highly likely that everyone will be wrong about something since so little information is provided. It is a great introduction to data science because, instead of looking at data represented with all the information, students are pushed to consider all the imaginable possibilities based on what they know and what they are curious about. This tool is designed to invite students into data conversations with an element of mystery that gets students invested as they await the big reveal. It is also one powerful way to bring together elements of science, mathematics, and social studies with real-world contexts to explore.

> **TIP**
>
> At the start of the day, show the first image for a Slow Reveal Graph. As the day goes on, reveal each successive image to share more information. Check in with students to see whether their predictions change as the day goes on. This could also be done across a week.

It has been argued that mathematics is fundamentally objective and exists independent of human interpretation. Many people believe that to be a math scholar, you must be socially awkward, wear large-rimmed glasses, and have wild hair. There is an existing narrative that mathematics is a solo sport, intended for an elite group that lives outside the realm of common practices and ways of being. As we've discussed earlier, mathematics has historically been used as a sorting mechanism in schools to determine who will follow the path that leads to superior positions in society and great wealth. Was math invented by math geniuses or does math exist even in the absence of humanity? Does the bee look to humans for guidance on designing the perfect hexagonal honeycomb? Do spiders study at the feet of scholars to determine the patterns that will use the least amount of silk yet trap their prey? Mathematics exists in nature, but has also been utilized to communicate the ideas that reflect the social needs and priorities of people. For far too long, mathematics has been presented as stiff or hard, having sharp corners. I believe that by softening those edges and finding the warm and fuzzy parts of mathematics, we can use it to connect mathematical ideas AND connect as humans.

Source: istock.com/FatCamera

Social studies is taught in elementary schools to build social understanding, develop civic competence, foster critical thinking, and guide students with integrating ideas. As students increase their knowledge of human societies, they seek to understand their place in the world and determine what type of citizen they will become. By studying and understanding history, students learn to attend to trends and practices that lead to societal evolution and strive to make a difference in the world. At every grade level, members of the community learn to embrace different aspects of social studies to make connections with others who may be the same or different from themselves. Does mathematics fit into the study of people? Can mathematics help us solve humanity problems? Can mathing be a social endeavor? "Yes!" "Yes!" and "Yes!"

As students begin their academic journeys, the first lessons they will learn is how to learn alongside others. Littles realize, sometimes for the first time, that their ideas are not the only ones that matter and that words can hurt but they can also heal. Children begin to see that they exist within communities, and in each different community they have a role to play and responsibilities for keeping themselves safe while respecting the needs of others. As they learn about their physical spaces, where to find the crayons or who they will sit next to each day, their exchanges with others begin to define a new reality to which they must adjust. From the beginning, mathematics will be a part of this new way of being because the number one lesson they will learn is that sharing is nonnegotiable. We share space, tools, resources, and talk time. Children adapt to a world in which their ideas are not the only ones that matter, and fairness is a core value that is budding.

It has been said that social justice doesn't belong in math classrooms. Some say that there are no connections between mathematics and activism. Others say that young children don't need to be introduced to advocacy work. But isn't that exactly what we should be using mathematics to do, solve real problems that are relevant to us? Everything that happens in elementary classrooms lays the foundation for the future work students will do in math class and life. Sometimes, we overthink things and make simple things complex when it isn't necessary. Have you ever seen a toddler who is falling apart because of an injustice? I mean losing it, having a full-on meltdown including tears streaming, snot running, mouth wide open, and uncontrollable wailing because they have been wronged. Whether "Sissy got more fish crackers than me!" or "It was my turn, but he kept playing!" Nothing will calm down this flailing little minion except justice. It started with mathematics because even young children who cannot count can determine more and less. And, mathematics will also be the solution. Get that baby more fish crackers! Isn't this the foundation of advocacy work? Yes, it is! Mathers advocate for fairness.

Source: istock.com/kool99

Advocacy work involves taking action to affect change by speaking out, educating others, securing resources for individuals or groups, and ultimately influencing decisions about unfair rules or policies. In early elementary grades, inequity and injustice look different than it will in the upper grades, but it is equally important and worth fighting for. By teaching students how to advocate for fairness when they are young, they will be equipped with tools to stand up for what is right when they are older. The third graders in my class had a reputation for taking a stand and positively influenced the culture at our school year after year. They understood the assignment and used mathematics to solve problems.

Here is one example of justice being served. Elementary students are excited to solve problems and become the hero in someone else's story. As teachers, we can offer this gift by giving students a purpose for "doing the math." One way to do this is to show students how math can help them advocate for themselves and others. When teaching third-grade students in Atlanta, our grade was assigned the last lunch block. Just before us, the fourth- and fifth-grade students had lunch. Each Thursday, and only on Thursdays, chocolate milks were available. There was a limited supply of chocolate milks, which by our lunchtime meant very few to go around. Week after week, my students complained about not getting chocolate milk, so I challenged them to find a solution. They immediately turned to mathematics under my guidance and with a few wonderings to probe their thinking.

Source: istock.com/Aletheia Shade

- I wonder how many chocolate milks are ordered each week.
- I wonder whether all of the students like chocolate milk.
- I wonder how many chocolate milks students are allowed to drink.
- I wonder how we could find out how much chocolate milk would be enough.
- I wonder who could help us solve this dilemma.

Even though it was not a formal assignment, my students were fully invested. They interviewed the lunchroom manager who was responsible for ordering milks. They created a survey for students to find out how many actually like chocolate milk. They worked together to determine how many chocolate milks should be ordered, with a few extra for good measure, and took their case to the principal. They had graphs, charts, and accurate calculations to demonstrate the problem and their proposed solution. Their diligence was rewarded with a solution that involved ordering more chocolate milks and limiting the number of chocolate milks each student could have. These third graders recognized math as a sense-making superpower and engaged in authentic advocacy work. Think about all of the skills that were practiced across content areas and how the work they did all year long with data influenced their approach to solving this problem. As third graders, they fought for fairness, and some day, as secondary students, they just might lead the charge when policies need to change. They have the tools, the experiences, and the confidence to stand up for what is right.

WHERE'S THE MATH IN THAT?

Entrepreneurs Are Mathers!

In an interview with Brittany Rhodes, mathematician, mom, tutor, education consultant, and entrepreneur, we explore the connections between entrepreneurship and mathematics. Brittany shares how mathematics can support students with having options for career choices later in life and how building confidence and developing a positive math identity just might be the key to finding success in business.

Deborah: What's your story? Tell us about your math journey.

Brittany: Math and I have been a thing for many years. I have always loved math and it has always been my favorite subject in school. I have NO working memory of not enjoying math. My mother was a career educator who knew she wanted to be a teacher at the age of seven, and by the time I was in high school, my mom became a founding principal at a K-8 charter school. So, education was a major priority in my household.

My mom signed me up for all types of programs growing up. Before STEM was a thing, I was a part of a program called Detroit Area Pre-College Engineering Program. I had many opportunities to participate in STEM activities and was introduced to STEM programs. I was learning HTML at 12 years old in Saturday classes, and eventually I started attending their 4-week summer programs, which were held on college campuses. This was my introduction to STEM, and I attended from 6-12 grades. By the time I was approaching my senior year in high school, I ventured to Atlanta and spent a summer attending a summer engineering program at Georgia Tech. I was fairly confident that I would go on to study engineering.

For college, I attended Spelman College in Atlanta and majored in computer science, but didn't enjoy it. When I decided to change majors instead of completing the dual degree engineering program, I wasn't sure what I wanted to do. My mom

(Continued)

(Continued)

	suggested that I major in mathematics, but I didn't understand how mathematics could be a major. I wasn't sure what someone with a math degree does. After exploring my options, I decided to major in mathematics. Since I had taken so much advanced mathematics in high school, I thought it would be fine. However, college mathematics kicked my butt. In the end, I stuck with it because I knew that mathematics teaches logic, reasoning, analytical skills, and other transferable skills, so I assumed that my skill set would be in high demand upon graduation. But when I graduated, the job market was not good because companies didn't see the connections between their fields and math.
Deborah:	I know that your business is focused on STEM, so if you didn't become an engineer how did you stay connected to the STEM field?
Brittany:	While I was a college student, I started tutoring other students in mathematics. That is when I realized that I was in the minority because it was obvious that many people did not like math at all. Tutoring introduced me to the concept of math anxiety, which I didn't know was a thing. It was while helping other college students that I fell in love with tutoring. I was great at helping other people see themselves as someone who could be good at math. With the job search being a bust, I decided to get a job as a tutor. To supplement my income, I did some marketing work on the side. My math brain started buzzing with questions, as I considered the role data analytics plays in market research and promotions. This led to me pursuing an MBA with a focus in Analytical Marketing to round out my pure mathematics background.
	I started to see the efforts around STEM becoming more popular, but math was in the background. There were volcanoes, coding, and robots, but it was like math was the appetizer, when it needed to be the entree. Math is not the background dancer. Math is Beyonce!
Deborah:	I agree! I've had many conversations about why we should capitalize the M in steM to emphasize that we should lead with mathematics to support the science, technology, and engineering.
Brittany:	I was working with students, friends, and family members tutoring in mathematics, but the demand would require me to clone myself. Since that was not an option, I started investigating new ways to make a greater impact with building math confidence. My husband, who I was dating at the time,

introduced me to the world of subscription boxes because he was subscribed to meal boxes. It was a new concept, but I was intrigued with the business model.

A year later, I attended a subscription box summit in Detroit. I thought my business was going to be focused on clothing, but I had an epiphany while on my honeymoon in Bali. I wanted to do subscription boxes to help kids with math. I called it Black Girl MATHgic because I like puns and because Black Girl Magic was trending thanks to CaShawn Thompson who had coined the phrase.

Deborah: That's fascinating! What made you decide to focus on girls in STEM?

Brittany: As I prepared to launch the business and quit my job, I started to read a lot of research that emphasized the absence of Black girls in STEM. Plenty of STEM boxes existed but none of them focused on math. I wanted to niche down further, so I decided to focus on girls since research revealed an urgent need. On my birthday that year, I announced to the world that I was offering a math specific box and Black Girl MATHgic was born! I wanted the box to be inviting to anyone, but I wanted Black girls to see themselves represented in this box as a learner and doer of mathematics.

It mattered to me that girls were entertained by and excited about the contents of the box, but I also wanted to make it meaningful. That is what inspired me to include the stories of real living and breathing Black women who were mathematicians. Girls needed to see women who enjoyed mathematics and were brilliant at their work that involved math. I knew that a part of the subscription would require me to find the women, interview them, and feature their stories each month.

Deborah: How did you determine who would be featured each month?

Brittany: Because of what had happened to me when looking for a job, a goal of mine was to prove that math majors have options other than becoming math teachers. The first box had to be amazing, as it was the launch of the business. So, I started with Dr. Gladys West, who pioneered the work that led to the GPS navigation system. The box theme was road trips, riding in cars, and the math we see on the road. I figured most girls could relate to riding in a car. We highlighted the math connections to road signs, mile markers, miles per hour, estimating distances, and more.

(Continued)

(Continued)

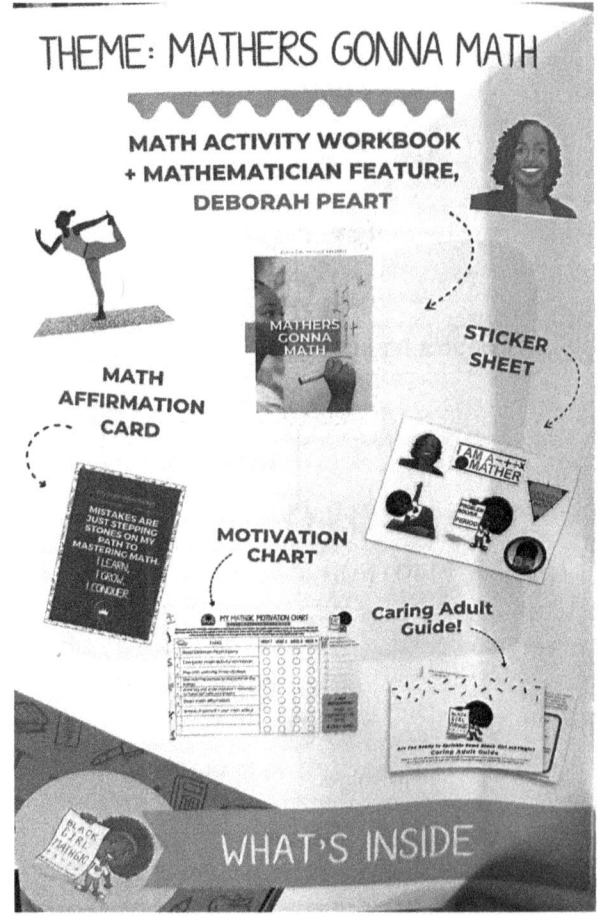

Source: Brittany Rhodes

Digging into the stories of Black women in mathematics made me realize that many of their stories were absent. The subscription boxes were well-received, and we did it for 5 years! It was very important to me that in addition to seeing math in the world, students were exposed to careers and their connections to mathematics. We curated about 60 boxes, including real estate agents, lawyers, doctors, fashion designers, sports agents, and of course we featured the wonderful Deborah, the Queen Mather too. The thing all of these women had in common was math.

Deborah: It was an honor to join the Black Girl MATHgic Mathematician Hall of Fame. Thanks for inviting me. What would you tell a student who believes they can be successful without mathematics?

Brittany: When students tell me they don't need math, the first thing I ask them is, "What do you want to do? What is your career aspiration?" Because of all the work that I did with curating the math boxes and interviewing the featured mathematicians, I could tell you how math is connected to almost any career. I learned very specific ways that math connected to our featured mathematicians' careers. So, pretty much anything a student says they want to be, I can give examples.

A favorite example of mine is when a student expressed that she didn't like math and she didn't need it to be an artist. That next morning in our session, I shared our box featuring Rachel and Rebecca Crouch. They are twin sisters who earned mathematics degrees from Howard University AND they are artists who own a thriving art business. They have even painted for the Obamas. And then I reminded her that as children, art often starts with lines and basic shapes. You guessed it, MATH is a foundation for art.

Deborah: How do we help teachers see the connections between their math mindsets and the impact it has on their students?

Brittany: We have a responsibility to help kids see math in the world. Teachers need experiences with connecting math to the world for themselves and they must believe that ALL students can be doers of mathematics. Math anxiety is an iceberg, and we are chipping away at it with a little ice pick. The middle school girls I work with have been forming their icebergs for 5-7 years, so chipping it away will take time. If elementary educators create positive math experiences and help students make connections between mathematics and their lives, maybe there won't be as many icebergs in middle school.

If we are going to open students' eyes to pathways and career options that require math, we need to ensure that students have dreams grounded in the reality that they will need math no matter what they want to be when they grow up.

Deborah: What else would you like to share about how your work impacts students' lives and learning?

Brittany: It is clear from the post-COVID data that math is not going well, and the kids are not all right. I wanted to find a way to impact more students, so we have transitioned to a new model working directly with schools, districts, and youth serving organizations. We've introduced a private brand label, Math = Me, but our focus is still on building confidence and decreasing math anxiety to support positive math identity development. Our box includes two important affirmations: "I Am a Math Person!" and "Strengthening the M in STEM!"

(Continued)

(Continued)

Math Equals Me

https://math
equalsme
.com

Source: Brittany Rhodes

> Now, I am speaking at conferences, offering professional development, math confidence workshops for students, and training to support the use of our kits in classrooms and for math intervention.

Deborah: Anything else?

Brittany: Whatever your reasoning for starting a business, you must know how much math is involved. You will not have a successful business without a foundation in basic mathematics because you are in business to make money. K–8th-grade math is the math you will be using when you are 20, 30, 40, 50, 60 years old. It makes me laugh sometimes because second-grade math shows up in my business every day. I just want to close by saying, Entrepreneurs are mathers!

Students who want to have a business one day may not realize that the math they are doing right now will matter. Of course, we don't need to tell an eight-year-old that they need to master multiplication because they may want to be an entrepreneur one day. On the contrary, we ensure that multiplication is presented in ways that are relevant to them now, so one day they have the option to start a business and the confidence to do just that.

TIME TO REFLECT AND TAKE ACTION

4

Sometimes we learn about new strategies or ways to hone our craft and immediately become overwhelmed because we want to make changes, but the reality is that we cannot do it all. So much is continually added to our plates, while nothing is removed. How can elementary educators be expected to stay in a continuous cycle of improvement without being given the time and space to make meaningful changes? We need to understand the difference between the things we can and cannot change. Then, we find the courage to change what is in our power to change and release the things that are outside of our control. Finding ways to connect across content areas is an investment that will yield big rewards.

1. What connections can you find between your science topics and math concepts?

2. Which routine will you try to support students with interpreting data regularly?

3. What is something that you can stop doing, so you can start doing something that will make a greater impact?

4. What are you already doing that is going well that you can adjust to get better results?

5. How will you support students with using mathematics to solve real problems?

MATHFIRMATION

Students need to know that math is a powerful tool that can be used to advocate for fairness. They should be encouraged to approach problems through an asset lens, which can empower them to persevere even when

(Continued)

(Continued)

math gets challenging. Remind students that they have everything they need to be successful with mathematics and that sometimes they have to encourage themselves. Self-talk is often a nagging nay-saying voice in our heads. Give students options to replace negative self-talk with positive mathfirmations. Have students practice saying them in a whisper voice and in their minds. There is power in having go-to responses when negative self-talk creeps in.

1. When you feel like you want to quit: I was born a mather and I have everything I need to succeed.

2. When the math is feeling tricky: Mathers make sense of problems and work hard to solve them.

3. When you fe
   the superhe
   figure this o

4. When you a
   Breathe out
   then maybe

5. When you f
   math. My b
   done.

5

DON'T SKIP THE FUN STUFF!

Source: istock.com/FatCamera

If there was a recipe for the perfect elementary school experience, I bet FUN would be the first ingredient. Children come to preschool bright eyed and bouncing with excitement. They come to learn, but they also come to play. The truth is, play is critical to learning and inspires curiosity, exploration, and creativity. The imaginary worlds children create while playing can become the stories they write. The lessons they learn from taking turns, listening to their friends, and remembering the rules of the game they've made up can translate into self-regulation skills and support character development. Even though we recognize that play **is** learning, as students get older opportunities to learn through play fade away. Sadly, play becomes a reward for hard work and good behavior instead of being a regular part of every day.

Across all content areas, students need unstructured and structured play, but in math class it is even more essential. Just as reading books of choice provide opportunities to practice reading skills and enjoy a fictional tale, and writing in journals allows for freedom of expression and practice with skills necessary to pen a great story, playing games in math class offers practice and application of skills learned and provides joyful math experiences. It is essential to include fun activities and free time for math games, but first let's disrupt the narrative that mathematics isn't meant to be enjoyable.

Take a moment to think about the tasks that are often skipped when time is short. Likely it is the activities that require extra materials, allow kids to move around or work in groups, and those that might get a little chaotic and noisy that we opt out of most often. In other words, we make time for the "hard work" and the "important tasks," but many of us are guilty of mostly skipping the FUN stuff. I know I was guilty of that until I decided more learning would happen if children actually looked forward to math class instead of dreading it. To be honest, once I made the shift, I myself looked forward to math class instead of dreading it.

I'm going to go out on a limb here and say that there is not a grade in elementary school in which mathematics has to be all serious and just about business. Honestly, mathematics can be playful well into adulthood if we explore more games and realize that math is often what makes them fun. Because of a mentality that we need to help students become serious about math, we rush them to let go of developmentally appropriate ways of thinking and shame them when they are not moving along at our pace. Yes, I meant that! The truth hurts, but the truth is we try to stay on pace and "nudge" children to hurry along toward more "efficient" strategies because they are taking too long to finish tasks. Maybe we applaud the strategies that we want all students to use and show apathy or judgment toward those we expect children to let go of by . . . (insert any age or grade level). I worked extremely hard to perfect my poker face, that face that doesn't give away which method I hoped to see versus those that I planned to strategically nudge students to let go of, in time. Students seek our validation, but giving it too often creates an insatiable need to gain approval before making decisions. For students to feel comfort and confidence in math class, they need to experience moments when decisions are theirs to make without our influence.

Students also need time and space to try new methods without being forced to adopt them. In our classroom, students learned to politely decline book recommendations and math strategies they deemed too challenging at the moment. A student might respond to a classmate who solved a problem in a different way, "That is cool and takes fewer steps, but I don't totally understand how you did it. I might try it later, but I'm not ready yet." Having this autonomy empowers students to take ownership of their learning. I intentionally created an environment where being honest about what you need or are ready for was respected. The students were kind to one another and encouraged classmates without embarrassing them. Diversity of thought was the norm, so they enjoyed hearing and sharing ideas without feeling pressure to abandon their own.

We adults could learn a few lessons from young children. We sometimes put pressure on children to find other ways to solve problems before they are ready. The number one example of this is "Finger Shaming." Young children intuitively use their fingers to help them solve problems and build their number sense, but are often ashamed to do it. They use their fingers to support them with computation, but discreetly place their hands under the table. Unfortunately, there is a belief that using your fingers is "for babies." Children are made to feel they are failing at math if they need their fingers because grown-ups make them feel that way. The irony is that we use a base-10 number system, so the BEST math tools available any time we need them are our fingers.

Source: istock.com/Kali9

> *There is a powerful connection between finger movements and brain functions like memory and concentration.*

Fingers can support students with early math skills, and using our fingers can enhance creativity and problem-solving. There is a powerful connection between finger Movements and brain functions like memory and concentration. Yes, you heard me. Wiggling fingers make neurons fire in multiple regions of the brain (Artemenko et al., 2022). Using our fingers to solve simple math problems or to count becomes a tactile experience and enhances numerical processing. Young children need to develop their fine motor skills, and doing so can support the development of early math skills. Not only should we avoid shaming young children for using their fingers, we should encourage practice using our fingers as a math tool. Intentionally supporting the use of fingers can help students with laying a foundation for long-term success in mathematics (Frey et al., 2024).

Recognizing finger patterns is an important foundational skill for combinations to five and combinations to 10. How convenient to have a base-10 math tool, literally at your fingertips! When using quick images for subitizing practice, we can incorporate 10-frames, dot arrangements, base-10 blocks, dice patterns, AND finger patterns. Students who can quickly recognize a quantity represented with fingers can become more facile with using their fingers for computation and move toward more sophisticated counting strategies (Wright et al., 2015). For example, students who know five fingers with automaticity can think about 5 + 4 as 5, 6, 7, 8, 9 using their fingers to count on instead of counting all. Eventually, they will move away from holding up five fingers to start and will just count on raising one finger at a time to reach nine. These foundational skills evolve into mental models and number flexibility, especially with visual representations and lots of practice.

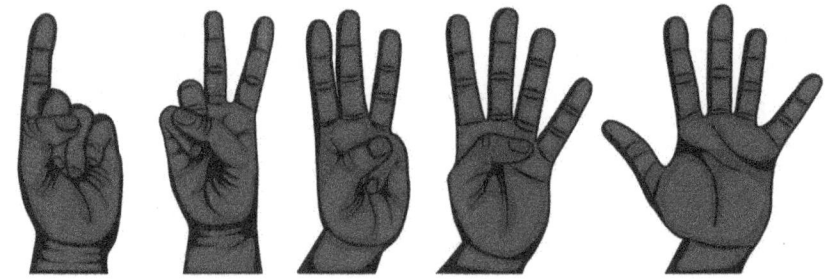

Source: istock.com/johavel

There are many ways to use fingers with young children, so they won't need to use them as they get older. The key is that we cannot rush the process. We can, however, help students make connections that will guide the transition when they are developmentally ready to move on. Have students make combinations of numbers in multiple ways using their fingers. For example, ask students to show five fingers, and then ask if they can show five fingers using two hands. Do this with any number within 10 to practice number combinations to 10. To make connections between number symbols, dot patterns, and fingers, hold up a card showing a numeral or dot arrangement and have students hold up their fingers to show the same amount. There are many partner games that students can play using their fingers too. Use games that match other representations and throw fingers into the mix. Be creative and find ways to encourage finger arithmetic because fingers are made for counting.

 READ. WRITE. MATH. CONNECT.

Tapping out letter sounds and clapping out syllables supports students as they develop phonemic awareness and syllabication skills. These practices are done publicly and sometimes in unison. We can incorporate finger use in math class to practice and develop counting skills and simple strategies so students recognize fingers as a helpful tool. This can help eliminate the negative stigma around finger use in math.

Eventually, it will be important to introduce abstract representations and simple arithmetic calculations. Fingers can support with this transition and help children grasp basic math concepts. When practicing basic addition and subtraction, have students put fingers up for adding to and put fingers down for subtracting from, which demonstrates how numbers can be composed and decomposed. Students who engage in finger-based arithmetic regularly can establish a solid foundation in basic number concepts. The irony is that students who are not finger shamed will eventually move away from using their fingers, while students who hide finger use tend to rely on their fingers longer because they miss out on the intentional practice to support the transition to more abstract numerical representations. We MUST leverage fingers as a math tool instead of just telling students not to use them. The truth is that many of us still use our fingers occasionally, so there is no shame in it.

> *Students who engage in finger-based arithmetic regularly can establish a solid foundation in basic number concepts.*

> ### ACTIVITY TO TRY
> *Friendly Fingers*
>
> During story time, have students hold up the number of fingers for people, animals, objects, or anything that can be counted as you read to mathematize stories and make them interactive.

Of course, we don't want students to rely on fingers forever, especially as quantities increase, but keep in mind that students might use fingers in new ways that align to more sophisticated strategies, even when reasoning multiplicatively. This is why we should ask students how they came to their solutions and how they used the math tool to support them. Fingers are no exception. When a student responds that they used their fingers to find the answer, ask follow-up questions like, "How did you use your fingers? Show me, please." Or, "Can you tell me what you said as you raised each finger?" Or simply say, "Tell me more about that." By making it OK to use our fingers publicly without shame or consequence, we free children to use the first math tool they discovered on their own. How liberating and validating to be celebrated for being resourceful!

So, the next time you find yourself unconsciously using your fingers to consider how long until the next available time slot or how many nights you need to book at a hotel, be not ashamed. Remember this moment, smile to yourself, and wiggle your fingers proudly, knowing that neurons fired and you made your brain light up! Give students the same advice and remind them that fingers are made for counting, and even grown-ups use them every now and then.

Mathematics has a place in the social studies block, but sometimes we miss the connections that can be made. Other times we don't realize that math is a part of the lesson. Especially in situations when there is a focus on words and acquiring new vocabulary, we don't always consider the mathematical thinking and reasoning that supports the understanding of these seemingly "math free" ideas. My students often questioned whether math was involved when tackling Perplexors (Gottstein, 1999), also known as logic puzzles, because the focus was on logic and reasoning. Deductive reasoning is based on rules or facts. By eliminating information that doesn't support your claim, using a logical chain of reasoning, the remaining facts help you prove a statement as true. We do this often in social studies and science, but this type of reasoning is also needed in math class. It may seem simple, but logic and reasoning are essential to establishing a foundation for making conjectures and proving that mathematical facts are true.

> **TIP**
>
> Offer students logic and reasoning challenges like Perplexors as math practice or extensions, and remind students that words belong in math class. Be intentional about pointing out how practicing skills like deductive reasoning will support them as Mathers.

Source: istock.com/Author

In social studies, we learn about maps and geography, starting with maps of our classrooms, schools, and communities. Students expand their vocabulary to include directional words and explore opposing positions in space. When children begin to use words like *up, down, in, out, over, under,* and *through,* these are words that have a fundamental connection to mathematics. Eventually, students contemplate the distance or relationship between objects in space using measurement, coordinates, and scale. Don't miss out on the opportunity to connect all the work that happens in social studies with maps, distances, and directional words to mathematics. There are also resources that combine these skills with math practice by having students solve simple arithmetic problems to find the coordinates on grids to color and reveal mystery pictures. Others help students practice geography

> **TIP**
>
> To practice with students, give verbal directions that involve moving forward, backward, left, right, up, down, etc., to get to a specific place in the room. For added enjoyment, use an instrumental version of any line dance or shuffle to get children up and moving. But don't forget to let them know that practicing direction words in very cool ways is mathing!

directions like north, south, east, and west, and subtraction practice at the same time. The final results are cool images and visual feedback that reveals you have solved the puzzle or cracked the code, while confirming you know your math facts.

Map reading and the algorithms that support GPS systems are grounded in mathematics, but we rarely show how the early experiences that lay the foundations for navigating our way through time and space are inherently mathematical. Understanding which way is up may seem inconsequential, but working on the coordinate plane in the upper grades will definitely require it.

Spatial awareness helps us recognize where we are in relationship to others and objects around us. Developing spatial sense in the early grades involves understanding how shapes fit together. To consider the size, position, Movement, and direction a shape is facing requires spatial sense. This comes more naturally for some students than others, but it is something that can be developed over time and with practice. Spatial sense is essential to understanding geometry, describing and classifying the physical world we live in, so honing these skills will serve students well.

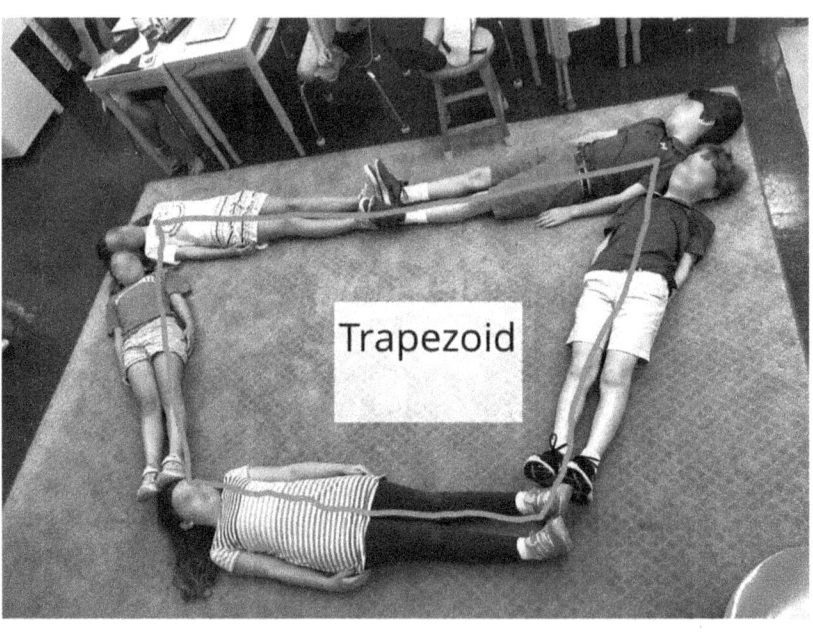

Geometry is a math topic that often gets glossed over, shortened, or skipped when we need more time. What we need to remember is that the simple skills taught in kindergarten will support future learning, even though it might not show up again until second grade, fourth grade, or later. So even if we don't see the immediate connections, introducing these foundational skills should not be skipped or minimized. Skills that are not considered major work of the grade might be found within supporting clusters or additional clusters in the standards, but they are still essential. It's important to remember that a house built on a shaky foundation will eventually crumble, so we must give students experiences that encompass them all. Young children need more than an introduction to identifying the shapes of blocks that perfectly show each defining attribute. They also need practice identifying shapes in everyday structures and objects in their environment. Shape scavenger hunts in books, around the classroom, or on school campuses support the basic understanding that shapes are everywhere and they matter. Recognizing that a triangle sitting on its base could support something but a triangle sitting on a corner or angle couldn't, is important. Highlighting the shapes in designs in carpets and ceilings, and noticing the shapes in furniture and buildings in class or for "homework," just might spark something artistic or plant the seeds within a budding architect.

> **TIP**
>
> Card games like SET and visual dexterity games like Q-Bitz are fun and develop spatial sense. Make these games available for times when there are a few minutes to spare or use them as a station when doing math rotations.

Geometry shouldn't be limited to the unit or a few days because there are so many natural ways to incorporate it all year long. Body-gons is an activity my students loved. Once students were familiar with a variety of polygons, I would look at the clock and say, "Looks like we have time for two rounds of body-gons." This was met with cheers every time! It is a silent activity, so the joy was expressed using the sign language for applause. Students quickly gathered around the carpet and sat quietly, as I pulled one name to start the round. The "builder" then pulled a card to reveal the shape that needed to be constructed and showed the group.

The builder carefully selected the number of students needed, considering heights, and began construction. The silent supporters pointed and waved offering assistance as needed. My job was to get a photo from an aerial view, meaning standing on a chair or desk and taken from above, so that the image could later be labeled by students. The rectangle presented an interesting challenge because it had four sides, but with only four students of varying heights, the group wasn't satisfied. They decided to build it with six students to make it clear that there were two long sides and two short ones.

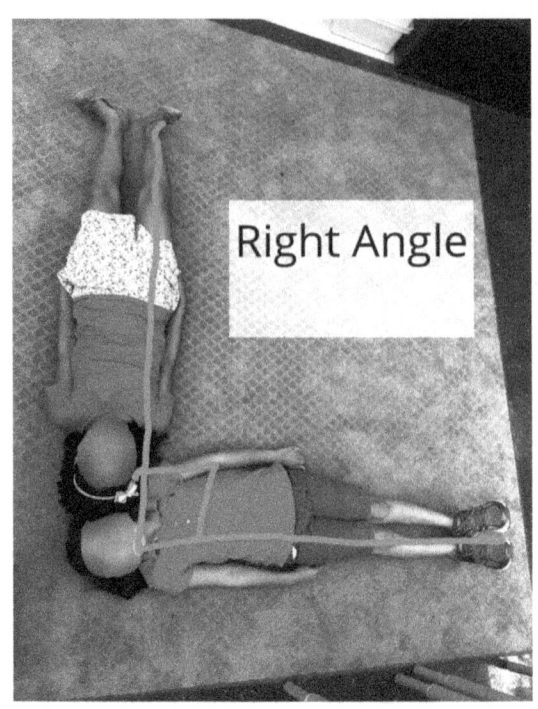

Another shape that required lots of input from the group was the hexagon. In trying to make a regular hexagon, they realized it was too big for the carpeted area. They thought it couldn't be done until one student motioned using their hands that two sides could be shifted to make a concave shape. They repositioned "the sides" and celebrated how they worked together to solve the problem before the time ran out. Sometimes on rainy days when recess was moved indoors, students asked if they could make body-gons, angles, or new shapes they had learned or wanted to try. Whenever possible, my response was a resounding, "Of course!"

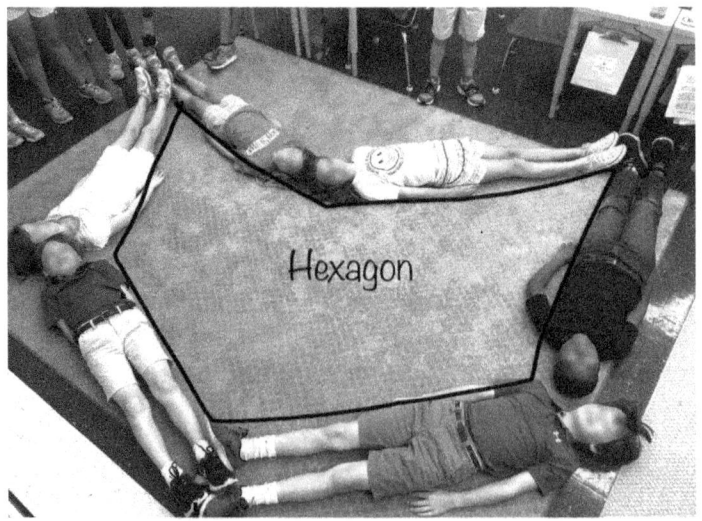

Mathnote
Geometric Mysteries

Geometry was believed to be divine according to the ancient Egyptians, and they are said to have had a deeply spiritual connection to geometry as a people. Because geometry shows up in nature in unexplainable ways, the ancient Egyptians believed the use of geometric shapes and designs brought balance and harmony to their lives. While they used geometry to map out the Nile River, survey land, and design buildings and temples, they also believed geometric symbols held mystical meanings. The Egyptians used the Golden Ratio, a mathematical concept found in nature, in their architectural designs and art. They incorporated geometry in their creation of hieroglyphics, included intricate geometric patterns in murals, and leaned on their understanding of triangles and angles to build the famous pyramids with precision. Ancient Egyptians used geometry in ways that have influenced science, mathematics, and art globally in immeasurable ways.

There is no shortage of games that students can play in school or at home to practice geometry skills. Qwirkle is a game that helps students practice matching colors and shapes, while also inviting problem-solving and strategy to the table. It is perfect for the classroom and family math nights because it is great for young children, while challenging older children to think strategically. There are different versions of the game available and even a travel set. It's also great for unstructured play. Children can create patterns or make up their own versions of the game. What is important is that students know that the fun games we introduce to them and encourage them to play with their family and friends are an extension of mathematics. Just as we remind them to practice reading skills by reading books independently or with someone else, remind them to practice math skills by playing games. Sometimes, it can be a wonderful option for homework instead of yet another worksheet or workbook page. Imagine the smiles and stories that will come up as students come to class ready to share how their "homework" went the night before.

Telling time is a skill that is essential to managing life, but with digital time pieces is it necessary to read an analog clock? The short answer is no, but there is value in using a clock as a math tool to make creative connections. A clock is a tool used to measure time, even though we don't often talk about it in that way. In life, we must learn to manage our time to schedule everything we need to do and want to do. In elementary school, this might mean becoming aware of the time of day when things happen or recognizing how much or little time has passed. A digital clock offers practice with reading numbers using a specific format, but an analog clock involves so much more. By having students interact with analog clocks, there are lessons to be learned related to time and other important mathematical ideas.

For example, the analog clock is useful for practice with counting by five, dividing circles into halves and quarters, and basic computation using the four operations. The structure of the clock is multilayered and quite complex. There are 60 minutes in an hour, represented by the numbers or tick marks around the clock. The "1" represents 1, but it also represents five minutes past the hour. The "6" represents 30 minutes or half past, and "12" represents o'clock, not 60. There are also 60 seconds in a minute, and the quick moving second hand, which is actually the third hand on the clock, stays in motion demonstrating that time never stands still. When the long hand is on the "3," 15 minutes have gone by since the top of the hour or it is a quarter past the hour because $\frac{1}{4}$ of 60 is 15. Now, the short hand points to the number that tells us the hour unless it is in between two numbers, which could

Digital clock   **Analog clock**

Source: istock.com/Udayakumar R

mean it's quarter 'til the hour of the number it's approaching. And then we ask students, "Do you know what time it is?" and expect that they could tell us, after the few days the "Telling Time" section of the unit lasts.

Source: istock.com/Venkatsh Selvarajan

The message here is simply, "SLOW DOWN!" Regardless of the curriculum you use, there isn't enough time spent on time (pun intended). Find ways to practice telling time all year long. Introduce the clock early in the year as a tool for measuring time. Even without formal instruction, valuable lessons can be learned from talking about how long something will take or when it will be time to leave the classroom to head to the gym or cafeteria. Ask time-related questions regularly, and point out when it is 11:00 a.m. or 1:00 p.m. Using timers that are visible to students, like large sand timers or a timer with large numbers, can help them begin to notice what one more minute feels like. When we did mindful moments so students could settle or transition, we always used a gentle timer to indicate the end of the time. One minute was challenging early on so we started with 30 seconds, but by the spring, third graders comfortably took time to breathe and refocus for three to five minutes. Sometimes they requested longer mindful moments after gym or when returning from an assembly.

The message here is simply, "SLOW DOWN!"

Another way to focus on time is to include start and stop times on the daily schedule. I had the schedule for each day printed and laminated without times to show the subjects and flow of the day, but each morning I wrote the actual schedule with times on the whiteboard. When we were running late or short on time, I would mention it and adjust the schedule as needed. Sometimes, I gave my students choices by letting them know we wouldn't have time for everything that was planned and asked which task they would prefer. Of course, this wasn't always an option but when it was, we discussed it and voted as a community, or village.

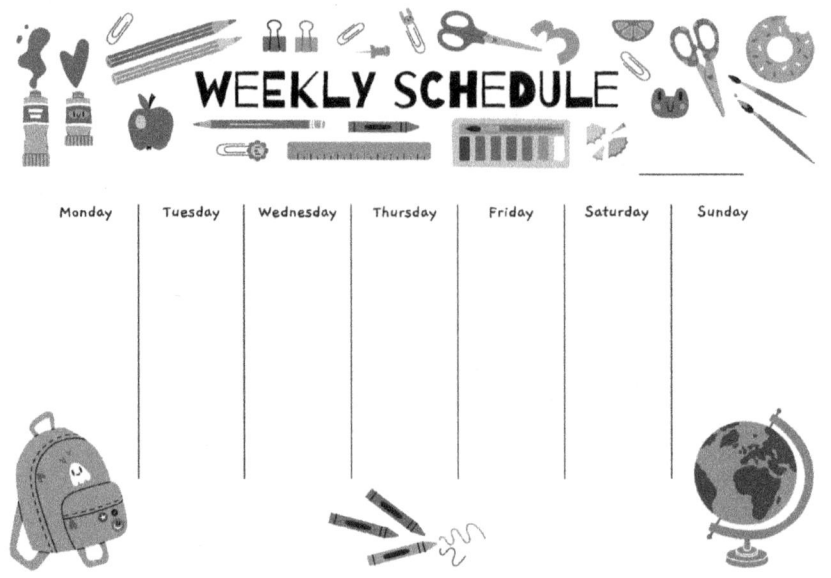

Source: istock.com/Elena Shiyukova

Whenever possible, we talked about the routines students had in their lives. When we did a survey about bedtimes, we graphed the results, and talked about the importance of rest for the body, especially the brain. Each student was tasked with coming up with a bedtime routine that could work for them with a family member. We also discussed morning routines, and how that might change on the weekend versus weekdays. During whole group read-alouds, we noticed the time of day based on the clues in the story. When possible, I also wrote story problems that included time based on the books we read.

When the unit that focused on time came along, students were beyond ready to read and label clocks, count by five, and use appropriate language to tell time. More importantly, they understood why telling time mattered and was relevant to their lives. By introducing the clock as a math tool for measuring time early in the year, the lessons on time didn't feel random. Students confidently learned new details about telling time and made connections between number lines and the tick marks and numbers on the clock. They recognized the difference between seconds, minutes, and hours, and understood that midnight and noon were written the same way, but one is 12:00 a.m., while the other is 12:00 p.m. Lessons on time went smoothly and they continued even after the unit ended because students knew that time wasn't just a few lessons for a few days of the year. Time was an ever-present part of our lives, and clocks helped to keep us all on track every day.

> **TIP**
>
> In addition to the regular clock in the classroom, buy two inexpensive clocks. Remove the minute hand on one, leaving only the hour hand, and remove the hour hand on the other. If there is a second hand, remove it from both clocks, and label each clock. By separating the hands this way, students get a better idea of how long it takes for an hour to pass versus how long it takes for minutes to pass.

When I was a little girl growing up in the Bronx, money was not just a few lessons that came up in the curriculum. Money was a tangible concept because if we had $1.00, we were "rich." Well, candy rich anyway. My friends and I would pool our money and go to the corner store to buy penny candies, which meant we could get 100 pieces of candy with a dollar. We pointed to Swedish Fish, Tootsie Rolls, Big Bols, Jolly Ranchers, Dubble Bubble, and more. Each of us counted along with the store clerk to make sure we got every piece of candy we paid for, and later we emptied our little brown paper bags and divided all the spoils equally. Some of my friends also used coins when they went with their families to the laundromat. On Fridays, we could purchase a slice of pizza, a bag of chips, and a soda for $1.25. Twinkies, Ding Dongs, cookies, or pies could all be purchased for 25 cents a pack. Doing chores and getting an allowance had immediate rewards. Money mattered to us, so we knew the difference between a penny, nickel, dime, and quarter. So, what is the incentive for our students to learn about money today? How do we make it relevant for them when they are probably used to seeing the adults in their lives pay for most things with plastic?

While the days of the penny candy stores are over, we can create meaningful activities to support students with understanding money. They may not need to make change, get coins for the arcade, or roll coins to take to the bank, but hosting a bake

Source: istock.com/Tamer Soliman

sale or having a class store might be a way to practice counting coins. Truthfully, we could probably skip lessons on coins based on the way children watch adults engage with shopping, mostly online, swiping cards in stores, or getting money out of the magic money machine known as the ATM. But, counting coins is beautifully aligned to counting, adding, subtracting, decomposing and composing numbers, and the base-10 system, which makes these lessons worth it.

> **TIP**
>
> Provide a money anchor chart for support that includes the value of each coin and bill. Set up your money station near the chart, so students can refer to it as needed. Change the coin collections in the jars periodically to give students lots of practice counting coins and finding the value of their collections.

Counting a collection of coins involves counting the number of coins and finding the value of a set. Working with values of 1, 5, 10, and 25 provides computation practice that is needed in the lower grades. By giving students story contexts like shopping, saving up for a special gift, earning money from chores, or planning for a school fundraiser, students begin to understand the give and take of money. When dollars are later introduced, the denominations are 1, 5, 10, 20, and 100, which are also perfect for basic computation practice and developing strategies based on place value for adding and subtracting within 1,000. Based on your school community and the communities represented by your students, find contexts that students can relate to that work well with building money sense.

Source: istock.com/ridvan_celik

Since money is often one of a few topics in a combined unit, this is another area in mathematics that I recommend stretching across the year. It is also one that could be enjoyable for students if we get creative. I introduced money early in the year and kept students practicing in a center activity I created called Money Jars. I purchased fake coins (and later fake bills) that looked and felt real and empty jars that were clear with black lids. The jars were labeled with letters. Each jar had a different number of coins and varying levels of difficulty based on coin combinations. Some students had jars with just pennies and nickels, while others had quarters, nickels, and dimes. Some jars had 10 coins, while others had 15 or 20. I had an answer key in a notebook, so I could differentiate by simply suggesting which jar a student should select by letter. They were NOT arranged in order of difficulty, so "A" wasn't the "easiest" jar. Students didn't feel they were at a low level or high level. It was just practice, and sometimes they worked with a partner.

On the sheet that I created to help students keep track of their thinking, Think Space, there was a place for students to put the letter of the jar, which made it easy for me to check and keep track of which jars students had completed. It also included guidance and reminders to support students with organizing the coins. Students shared the number of each coin, the combined value of the coins, and the strategies they used to find the value of the set. Students then recorded an explanation of their solving methods, using Seesaw, and uploaded it for review. Every few weeks, I emptied the jars and created new coin collections. Later in the year, I included bills in some.

 ## ACTIVITY TO TRY
Money Jars

Create your own Money Jars station or center, so students can practice counting money all year long. Use clear jars (plastic baggies can work but aren't as durable) labeled with letters on top. Include coins that students have already learned the value of in a variety of combinations. As students are introduced to more coins and eventually bills, swap out the jars to make them more challenging. Make sure students have a way to keep track of the number of coins and the value of the set, and provide space for them to show their thinking. These jars are meant to have a limited number of coins, so students can practice when there isn't a lot of time. One way to extend this activity is to use coins for counting collections and have students work in groups. Students can gain practice counting larger quantities, share grouping strategies, make connections to base-10, and calculate the value of their collections when they are ready.

Source: istock.com/AfricanImages

Students gained valuable computation practice using the context of money. Once they recognized the coins and knew the value of each of them, they moved through money tasks with ease. I assigned values to pattern blocks and had students create people and objects. Then they calculated the total value of their creations. Sometimes, they were tasked with creating something using the blocks with the caveat that it had to be "worth" a given amount.

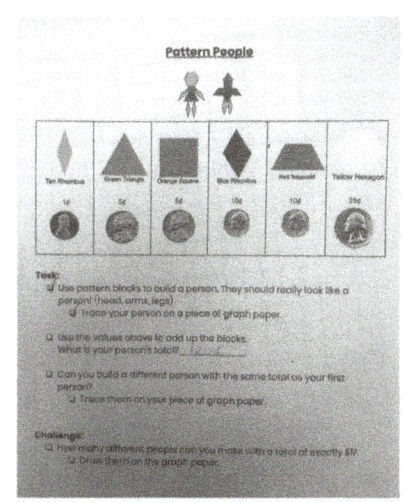

In one of my favorite activities, my fourth graders had to plan a party with a specific budget, but they had to get everything on the provided planning list. When presented with options like buying individual party favors or buying them in bulk, they determined the best deal as a team. To make the experience interactive, I created a slide presentation that was set up as a shopping trip with different stops along the way. As I clicked through slides, we "left" stores and "arrived" at new ones. Each group made decisions in real time while managing their budget, subtracting as they spent money, and checking off items on their shopping list. All year, students had opportunities to flex their money muscles and sharpen their computation skills with periodic group challenges or independent practice. We had a blast every time!

 READ. WRITE. MATH. CONNECT.

In addition to including Money Jars as a station or center year-round and creating tasks for students to apply their money math skills, we can write story problems that include contexts with money to practice computation skills in meaningful ways. These problems can be infused into regular assignments as opportunities for practice. As students become facile with money calculations, they can write story problems that include saving money, spending money, counting and comparing money, and more. There are life lessons and financial literacy skills that can be integrated, even in the early grades.

Across the year, there are many topics that must be covered, and time always seems to get away from us. Unfortunately, this usually means we skip the fun stuff. Of course, we can make just about any math topic fun, but sometimes those math concepts represented by the additional and supporting standards are viewed as "just for fun" when they actually teach essential skills and provide contexts for application of core math skills. Let's be proactive and plan the ways in which we will incorporate the fun stuff all year long. If we focus on ways to make it meaningful and relevant for students, these topics can provide opportunities for applying the skills from core math content and be a source of joyful experiences. The other benefit is that students will learn to associate having a good time with friends with mathing. Instead of only using practice pages, we can find playful ways for students to reinforce and strengthen their understanding of the major work of the grade. I'm also willing to bet that the grown-ups in these spaces will experience a bit of math joy of their own, as they watch students work hard and play harder mathing in a community.

> Let's be proactive and plan the ways in which we will incorporate the fun stuff all year long.

WHERE'S THE MATH IN THAT?
Triple Threats Are Mathers!

In an interview with Rose Jackson Moye, singer, dancer, actress, and teacher, we explore the connections between the performing arts and mathematics. Rose shares her story of a young girl who literally marched to the beat of her own drum and danced everywhere she went. As a triple threat, Rose shares how when mathematics was relevant and applicable it supported her successful career as a performing artist.

Deborah: Tell us about what type of student you were growing up.

Rose: I am the daughter of a teacher. My mom was big on education and believed it was the cornerstone of EVERYTHING. We had a chalkboard in our kitchen. However, me and school, just no! I loved going to school to be social and only did what I had to do, so I could do what I wanted to do. Elementary school was laughable because I switched schools every year. My mom wasn't satisfied with the level of teaching, so eventually she got me into the school where she worked. That is when she realized it wasn't the schools or the teachers, it was the girl. She noticed that I spent more time stopping by her classroom to say hello than sitting in class. On standardized tests, I filled in the bubbles to make pictures. My mother realized she needed help.

My mom took me to a psychologist when I was around 8 years old for testing and discovered that I had dyslexia. The psychologist asked, "Why are you trying to force this round peg into a square hole?" The psychologist cautioned my mom that if she didn't acknowledge that I needed to learn things in a different way, mostly through play and hands-on experiences, she would be setting us both up for a dreadful education journey.

High school was a blast because I was a cheerleader, I played in the band, and I danced. Everything that happened between the hours of 8 a.m. and 3 p.m. were a blur, but everything that happened after-school was awesome. I made Bs and Cs in school but there were no supports in place for students with learning differences at that time, so my mom signed me up for a speed-reading class. It didn't "fix me," but it helped. My mom supported me throughout my academic career, even when she realized she was paying college tuition for me to dance.

Deborah: What was your relationship like with math?

Rose: I didn't feel anything towards math. It was just a regular part of my life, so I just did it. My dad had a cab company, which my mom ran. On Saturday mornings, cab drivers had to pay their rent for the dispatch service. So, Saturday evenings were spent counting all the money on our kitchen table. My dad was a stickler, so at a very young age, I had to count money, organize the bills, and band them. And then, we had to cross-reference the books with the actual cash. I don't know that I realized what we did as a family every Saturday was math. I was just super comfortable with numbers, so I felt nothing about math per se.

Now, in high school, when they started replacing numbers with letters, I was not okay. I thought that it was a personal vendetta against me that letters were showing up in math. It didn't make any sense to me. I didn't think this type of math was relevant to me because I just needed to count money and count measures in songs. I didn't see the connection to my life, so I asked my mom

(Continued)

(Continued)

to advocate for me. I believe that conversation we had with the psychologist all those years ago is the reason why she didn't fight me on it. I opted out of math at the end of tenth grade and took extra classes in the arts. I knew I needed math, but I didn't need THAT math. Unfortunately, opting out of math in high school landed me in a remedial math class in college.

If I didn't want to take the math class in high school, I wasn't sure why they thought I wanted to try again in college. As a matter of fact, I danced in the Bayou Classic as a solo dancer with the marching band because instead of being in that remedial math class, I was in the band room advocating for an opportunity to improve the dance routines. The crazy thing is that the formations on the field required more math than I would have gotten in that class. Sometimes the formation was the half-time score, which is different for every game and unknown at the start of the game. To be in the marching band required musicianship, but it also required precise math skills. The graph that showed the routine and formations was intense, but it made sense to me. If I was off when I came down the 50-yard line, the trombone might smack me in the head.

Deborah: Now, that's some math that matters! That math had immediate consequences. I'll bet no one invited you to come to the band room to do some math.

Rose: I think if someone had called it math, it may have landed differently, but because no one did I was totally comfortable.

Deborah: Tell us a little about your career as a singer, dancer, actress, and teacher.

Rose: When I was a student at the New Orleans Center for the Performing Arts, a teacher sent me to an audition at Opera Land, which was like Disney World but with Broadway-level performances. That was the first time I had to put singing and acting with dancing, so this was the experience that prepared me for NYC and a job on a cruise line. When I auditioned for the understudy of Carrie in *Carrie: The Musical*, based on Stephen King's novel, on Broadway, I was comfortable with learning my part that was mostly dance and the lead role that was mostly singing. From there I toured with *Ain't Misbehavin* and continued to land roles.

This journey made it clear that I was my brand, so I needed to sing, dance, and act if I wanted a successful career and to make a good living. My comfort level, my inability to hear "No," my determination, and flexibility are the reasons why my resume is long and my career was so robust.

Deborah: Wow! That is all so fascinating. I know you mentioned stage directions and choreography required math you didn't think about, so it was easy. What do you think it takes to be good at math?

Rose: Math wasn't the issue for me. I was great at math when the math applied to my interests. I don't think it ever was about being good at math. It was more about math being relevant and interesting. When it was, I applied myself and it went well. The mathematics that is a part of musicality might challenge someone else, but I was right at home. With choreography, you have to consider that musicians count fours and dancers count eights. You need to be aware of the number of people on the stage, the spacing, and the timing, which are all mathematical. At the end of the day, the math matters with the spacing and the placement of the dancers because if you don't get the math right, there will be chaos but when you do it's a beautiful performance. None of this was called math, but it is definitely math. It's quite complex actually, now that I am thinking about it.

Deborah: What would you tell a student who believes they can't be good at math?

Rose: First of all, mindset is everything! In the dance studio, my students knew that "Can't" was not an option. Having a belief that you can't do something, is not an option. So, math wouldn't be any different. I believe there is not anything that you can't work hard to accomplish. So, if you can dance, you can math. When things get hard, we keep going.

Deborah: Is there anything else you'd like to share?

Rose: Math is the study of patterns, so it makes sense to me now that real-world math felt natural to me. If it wasn't for the kitchen table math, with no calculators I might add, my level of comfort with numbers maybe would have been different. In the band room, I was at home because I just felt good with numbers in general. I actually find numbers to be Godly, like the phases of the moon, the calendar, and all the math we find in nature. The systems, the cycles, and the patterns that numbers provide, work for me. If students had more opportunities to investigate patterns, they would probably feel better about math.

Well, there's our call to action: "If students had more opportunities to investigate patterns, they would probably feel better about math." Whether students want to be singers, dancers, or actors, they need math. Our job is to help them see the connections between mathematics and the things they

(Continued)

(Continued)

enjoy. What a gift to feel at ease in spaces where complex mathematics is happening. There can no longer be a mismatch between "classroom math" and real math. If we want to build a community of mathers, we need the math to always be mathing!

Source: istock.com/ChrisGorgi

5 TIME TO REFLECT AND TAKE ACTION

Sometimes we need to get creative to manage our time and get it all in. In math class, this often means cutting out the "chaos" and flinging out the fun because the fun stuff takes too much time. It's time we think outside the box to find ways to cover the standards AND have a good time. We need to shift our mindsets to embrace play as essential to math learning and establish a standard of excellence that includes hard work and math joy. Additionally, we need to adjust our attitudes about who deserves to engage in fun math activities. The correct answer is ALL students deserve to have joyful math experiences. Games are not meant to be used only as a reward, and sometimes we need to skip the boring practice and replace it with something more inviting. This will take time on the front end, but it will be well worth it.

1. What resources will you leverage to find new ways to approach your least favorite math topics?

2. How will you incorporate practice for time and money outside of the math block?

3. Which books do you have that align to math topics? How can you use them to extend math practice?

4. Which childhood experiences of yours can you look to for math inspiration?

5. How will you support students and their families with connecting mathematics to the games they play?

MATHFIRMATION

As students learn to recognize mathematics in areas of their lives that were previously believed to be absent of mathematics, they will need ways to maintain their positive attitudes. It is important not to overwhelm students by constantly saying, "This is math!" but they do need to see the connections to mathematics to overcome fears and anxiety associated with the subject. To help students shift their perspectives, do periodic check-ins using the sentence frames that follow. Make it a regular practice to check in on developing math identities.

1. I used to thi
2. Math used to
3. Did you kno
4. My favorite

PART 3

BUILDING A COMMUNITY OF READERS, WRITERS, AND MATHERS

Source: istock.com/monkeybusinessimages

6

BECOMING THE MATHER YOUR STUDENTS NEED YOU TO BE

Source: istock.com/Fatcamera

When it is time for reading, students hear the enthusiasm in my voice: "Join me at the carpet, so we can see what Dyamonde Daniel is up to today." The children in my class know there will be animated voices, and our discussion will be rich. During the writing block, I open by sharing what I have written based on the prompt that I have given them, modeling the literary devices we are exploring. When my students read, I read. And when they write, I write. Even when assigned writing projects based on books, I often participate to show them how invested I am in their experience. Yet, math class starts about the same each time: "Please get out your math books and turn to page . . . " There is no invitation. Children prepare to join me in the chore of plugging away at the math of the day. I didn't hate math, but it was obvious that it wasn't my favorite subject to teach. Honestly, it hadn't occurred to me that I could make it more interesting because we had to follow the curriculum lockstep, right?

At every school where I have taught, there were clear expectations about mathematics instruction. Reading and writing were celebrated with bulletin boards showing off the latest projects and creative displays of a job well done. Math, on the other hand, was like our dirty little secret that lived within our notebooks. Teachers were expected to maintain a different level of order during math class. There was no room for collaboration and discourse or special projects. Mathematics was very serious and warranted extreme focus. Perhaps that is why you could hear a pin drop during math class, unlike other times of the day when laughter erupted or students shared their opinions passionately. Any math clubs that had any hint of exploration were reserved for a select few. The crazy thing is, they didn't feel like it was a privilege, but more of a punishment. For their great "talent" in mathematics, students were rewarded with more mathematics. I began to notice that whether students were "good at it" or not, there was "NO Love" for math. In my effort to be the best teacher I could be, I knew something had to change.

Source: istock.com/damircurdic

We all know that an anchor is a device that is used to keep a vessel from drifting. It provides resistance against the natural forces that would cause the boat to float away without clear direction. The anchor is grounded and provides stability for everyone on board.

Source: istock.com/artversion

When teaching about mindfulness with educators, I use this example to demonstrate the importance of taking the time to center ourselves before guiding our students because we cannot help them settle if we are riddled with tension. I found that this was also true for shifting my math teaching practices. To be more enthusiastic and invested, I needed to be genuinely interested and excited about the content. It occurred to me that I didn't enjoy learning math as a student, even though I made excellent grades. There were no fond memories to tap into like I had with reading and writing, so I had to find ways to make math more interesting but, in my heart, I didn't believe it was.

I remembered when I was a student myself being told to stop asking questions in math class, so I believed curiosity didn't belong there. I reflected on trying to understand why an algorithm worked and having teachers chastise me for being sassy or rude, so I thought mathematics wasn't supposed to make sense. Math class was a place for following rules and getting correct answers, which required little to no emotion. How could I possibly bring this subject to life for my students when I felt nothing?

My answers came when I took a math methods course as an elective at the end of my master's program. For the first time, I began to understand the *why* of mathematics and not only the *how*. My view of mathematics was forever changed because I had experienced mathematics in a new way and realized that I had been robbed as a student. I questioned, "Why didn't my teachers explain this to me?" Later I recognized it was likely because they taught as they had learned, so they probably didn't know the answers to my questions.

Source: istock.com/Olga Ubirailo

But once I turned around my thinking as a teacher and I began to understand number relationships and the underpinnings of the math concepts I knew on a surface level, I couldn't contain my excitement. I urgently wanted my students to experience this feeling and develop a positive relationship with mathematics. It was only the beginning, but I was on a journey to become the math anchor for my students.

 READ. WRITE. MATH. CONNECT.

> We often provide an anchor for our students in reading and writing without even realizing it. Our love for reading and our comfort with reading and writing shine through and provide inspiration for students. Many students aspire to be better readers and writers because of the behaviors modeled by adults in their lives. That is what it looks like to be an anchor for students. We read to them and with them, encourage them to revise and improve their brilliant ideas in writing, and smile as we reassure them that their hard work will pay off. Let's provide the same level of stability in math class by being an anchor for them, so our students grow as mathers.

As the anchor, I needed to present mathematics in a way that would engage students and keep their attention from drifting away. I needed to provide resistance against the narratives that tell students they aren't capable and the stereotypes that deem some students unworthy. It was my responsibility to be grounded, even while still learning, so I could provide stability for everyone on board. I began reading everything I could and signing up for classes, workshops, and conferences. Before long, I was championing mathematics in our building and leading the charge to create a math culture of sense-making and belonging for all.

The real work began with my colleagues. In an effort to improve math experiences for students, we had to rewrite our own math narratives. Some needed to heal from math trauma, while others needed to manage their own math anxiety. We all needed a mindset shift and to sit in the seat of the learner, so we could approach mathematics through a new lens, one that included joy.

Source: istock.com/SolStock

Eventually, things began to change.

Comfort B4 Confidence: A Framework for Change

> **TIP**
>
> Ask your colleagues for advice or feedback about a task or lesson that did **not** go well to get a conversation going. Be open to suggestions to try a new strategy for student engagement. Let's normalize being vulnerable to hone our craft.

In my current work with elementary educators, I develop professional learning using the Comfort B4 Confidence Framework. It is grounded in the idea that it is necessary for teachers to develop a positive math identity themselves in order to nurture positive math identities in their students. I aim to help participants build comfort by having positive and joyful math experiences before approaching new concepts and instructional practices.

There are 5 Cs (Comfort, Competence, Confidence, Collaboration, and Community) that outline the process I went through when shifting my beliefs about mathematics and taking the leap to advance my instructional approach.

As my

- Comfort with mathematics improved, I studied and developed my
- Competence and grew in
- Confidence. Once I felt grounded in my approach, I was ready to
- Collaborate with others and build
- Community.

Year after year, the math culture in my classroom became visibly positive, and students were more and more enthusiastic about math. My colleagues observed often, and were inspired to make changes in their own classrooms. Over the next decade, through reflection and intentional design, the Comfort B4 Confidence Framework was established as I supported others with becoming the mathers their students needed them to be.

The 5 Cs are not interchangeable, but the process is cyclical. As we grow and change, as we move from one concept we must teach to the next, we find ourselves having varied levels of comfort. **Comfort** must precede building pedagogical content knowledge or **Competence**. Only when we understand the math we teach conceptually, can we gain a greater level of **Confidence** in our ability to teach mathematics in engaging and meaningful ways. Once teachers have a positive mathematical identity—defined by Aguirre and colleagues (2013) as the dispositions and deeply held beliefs students develop about their ability to participate and perform effectively in mathematical contexts and to use mathematics in powerful ways across the contexts of their lives—they can willingly **Collaborate** with others and build a **Community** of mathers.

> *Comfort must precede building pedagogical content knowledge or Competence.*

Too often, we as teachers are expected to implement new curricula or instructional practices without the necessary training and support to help us feel comfortable. The first criteria for building comfort is that teachers need to do the math **for** themselves. Many professional learning sessions are focused on taking in information to immediately think about how it will be used with students. Of course, this is one goal of professional learning but we don't have to get there as quickly as we think. Especially with mathematics evoking an emotional response for lots of people, educators need time to process new information, try out their strategies, and make connections. Change can be hard, but with time and space to truly understand the math we teach it can be a lot easier. It also helps if there are low stakes and a level playing field, so trying out new methods in an evaluative environment will not work. Building comfort requires vulnerability, so we need to create spaces for teachers to explore and discover new approaches to solving problems with their peers. Teachers need to know they will be respected and that meaningful conversations will lead to lasting impact and professional growth.

Professional Learning Communities (PLCs) were meant to be spaces for learning together, analyzing student work to gain insights about student thinking, and shifting practices based on the things we have learned. Unfortunately, in many places, PLCs devolved and became meetings to discuss Pacing, Logistics, and the Calendar. If we reclaim PLCs to honor the learning that should take place, one small change that can make a big impact on building comfort among teachers is starting every meeting with a math warm-up routine. Doing math together is the first step toward building collective comfort and shifting mindsets and beliefs.

> *Doing math together is the first step toward building collective comfort and shifting mindsets and beliefs.*

What do you notice? What do you wonder?

Warm-ups provide a safe option because they often have multiple paths for solving and in some cases multiple solutions. Many curricula have some form of math warm-ups, so teachers can try out some of the routines from their curriculum with each other before leading

Steve Wyborney's Esti-Mysteries

https://qrs.ly/t8gnmto

them with students. There are also many options online like Steve Wyborney's (2022) Esti-Mysteries and Estimation Clipboards, which I highly recommend for practicing estimation and occasionally good belly laughs. Even teachers who are initially uncomfortable will recognize the low stakes involved and eventually participate willingly, maybe even looking forward to them. Ideally, everyone can take a turn leading a warm-up, but early on the teacher advocating for change will likely need to get the ball rolling.

Because many of us learned math differently than we are expected to teach it, it is critical that we retrain our brains to approach math in new ways. Although math is not new, we need more effective approaches to better support students. If we want our students to think flexibly and not rely on rote memorization or procedures that won't always serve us well in the long run, we need to try out new methods for ourselves. Particularly with mental math strategies, we fall short because we have limited math thinking flexibility. But, with practice we can improve.

When adding or subtracting multidigit numbers mentally, if your go-to strategy is to "stack the numbers in your head" we have work to do. Doing number talks or number strings with colleagues can help us see other ways of thinking and stretch us. Hearing others explain how they thought about a problem can unlock the mather within, so we can begin to trust ourselves a little more each time. Remember, we were all born mathers with math intuition and an affinity for pattern seeking. So really, this isn't about learning something brand new; it's about reconnecting with a part of ourselves that has been suppressed. Over time, as we rewire our brains and neurons fire and form strong connections, we build our capacity for solving math problems mentally and eventually build comfort.

> Because many of us learned math differently than we are expected to teach it, it is critical that we retrain our brains to approach math in new ways.

At the risk of sounding like a total nerd, I offer one last suggestion for building collective comfort. It is something that we did at our school that changed everything. A colleague and I started hosting math parties. Yes, you heard right, small gatherings with food and beverages where we solved math problems together. I don't mean super challenging college-level problems. As a matter of fact, we started with fourth- and fifth-grade problems because the new curriculum we had adopted was vastly different from the previous one. The events were optional and took place on a Thursday or Friday evening. Only a few joined us the first couple of times, but eventually it became a community event.

Math parties were both humbling and inspiring. We were all working with a new curriculum, so it was definitely a level playing field. We challenged one another to resist the urge to use the algorithm right away and instead approached each problem as the curriculum instructed. We pulled out math manipulatives, drew on graph paper, and compared our work often. There were times when we all got the problem incorrect, but in discussing and comparing our strategies we figured out the correct path. At other times we all had the correct answer, but our pathways to the answers were totally different. The beauty of it all was that we were openly sharing our mistakes, learning from them, and growing in our comfort with the new curriculum.

Our students benefited from our productive struggle, but in this setting, they were not our focus. We gave ourselves permission to be learners. We fought for our own understanding, supported one another, and laughed together. This spilled over into our classrooms and began to shift the culture because we talked about math at lunch and in the halls. We comfortably shared our successes and failures with teaching lessons and strategized to shift our practice together. Our students witnessed their teachers asking questions

> We as teachers had a level of comfort with mathematics that liberated us from pretending to know everything.

of one another when math didn't make sense and soon followed our lead. We as teachers had a level of comfort with mathematics that liberated us from pretending to know everything. We established a norm of admitting when something is confusing, and made it clear that math belonged to us all. In building our own comfort, we helped our students to do the same.

Building Competence: Understanding the Math We Teach

Have you ever asked a student to explain their thinking, but after several tries you still aren't sure what they did? Even worse, you have no idea why it's working. In my earlier days of teaching I found myself discouraging students from using inventive strategies and robbed them of opportunities to make conjectures and test them because of my lack of conceptual understanding. Even when students were stuck in the middle of grappling with a problem, it was a challenge for me to do something other than tell them the answer because of my limited exposure to multiple solution paths. It occurred to me that I was limiting my students' understanding and disrupting sense-making by introducing algorithms and short cuts prematurely, or over-scaffolding and spoon-feeding next steps. Harris (2025) is correct in saying that we must "avoid the trap of algorithms" because it can be detrimental to our students' development of mathematical reasoning. As for me, I felt more comfortable with teaching math, but something was still lacking.

Building comfort is necessary to dig into math curriculum and plan to teach a lesson, but building competence is required for conceptual understanding. With a new level of comfort, I could successfully execute each lesson and confirm it went well based on the teachers' guide that was provided with our curriculum. Students were engaged and didn't hate math as much, but I knew this was only the tip of the iceberg. As long as students asked the questions that could be answered by the answer key and teacher notes, everything was perfect. But there were those questions, the ones that often started with "Why . . . " that I evaded because something was still missing. I realized that I wanted more, so I went back to graduate school to learn more about elementary mathematics.

READ. WRITE. MATH. CONNECT.

Book studies are a wonderful way to explore topics of interest, shift mindsets, and improve instruction. We have often participated in book studies to enrich our experiences and refine teaching practices across content areas, but mathematics hasn't typically been the focus. Consider leading a book study based on math concepts, math teaching practices, or a specific instructional routine. It is encouraging to try new things with colleagues and friends and share strengths and stretches. A book study can be one step in the right direction of developing collective agency and shifting the math culture in your community.

The math methods course I took sparked something in me and piqued my interest, but taking courses focused on understanding how children develop number sense, the power of conjectures in elementary school, math progressions, and, most importantly, a conceptual understanding of mathematical ideas, changed the game for me. There is a misconception that elementary mathematics is simple and easy, so anyone can teach it, but that couldn't be further from the truth. There is nothing elementary about the foundational mathematical skills students learn in the early grades. What I found was that in building up my competence, I was better equipped to ask the right questions to keep students thinking and striving instead of rescuing them. This is why all elementary educators need to understand the math we teach, conceptually. Although we can't all go back to school to take additional courses, we can invest the time to learn more about early mathematics development, if only in small bites.

Source: istock.com/andresr

> *There is nothing elementary about the foundational mathematical skills students learn in the early grades.*

TIP

Keep questions on a note card that you can ask students to push them in their thinking before offering additional supports. Prepare scaffolds in advance for the times when students get really stuck, but do not offer them prematurely. Encourage partner work for finding solutions to tricky problems so students don't wait with a raised hand to be rescued.

Learning happens when children struggle productively and make mistakes along the way, but if we don't know how to guide them toward deep conceptual understanding and encourage them to persist, they can become overwhelmed and seek out the path of least resistance. Building our competence as elementary math teachers is essential to our professional growth and critical to student success. Whether we make time to read professional resources, sign up for webinars or workshops, or take an online class, it is important to find ways to build pedagogical content knowledge. Unfortunately, most undergraduate and graduate programs don't adequately prepare elementary educators to teach mathematics. While teachers don't need one more thing on their plates, adequate preparation is an urgent need. It requires a level of investment that seems unreasonable, but I can attest that it is well worth it and makes a world of difference.

Not only do the national math scores demonstrate the math crisis that currently exists, but the demand for problem solvers in the workforce continues to go unmet. Math class is a place where critical thinking and strategizing can be applied and skills we need to function in society can be sharpened. We as teachers miss out on opportunities to integrate mathematics across content areas when our pedagogical content knowledge isn't strong, and, as a result, the students experience a lackluster version of what mathematics should be. If we want to improve student achievement in mathematics and shift mindsets and beliefs about what it means to be "good at math," we need to step up our game. Competent math teachers become confident math teachers!

Building Confidence: Thinking Flexibly and Affecting Change

We teachers are excellent at convincing students to love something they previously despised. I witnessed many students falling in love with reading after a stretch of faking it for the sake of earning stickers. Something happened that made them want to learn to read, want

to find out what happens next in the story, or want to escape reality if only for an hour. At the start of every school year, I was confident that my students would be competent readers and that many would fall in love with reading. Because of my love for reading, my competence due to my extensive training and ongoing professional development, and my drive to ensure the success of my students, it was not an option for a student to leave my class at the end of the year the same as they had started. Having been in my class had to mean something, and growth was the goal for all. I was determined to put in the work to help each student improve as a reader. After reflection, I realize that one of my greatest assets to the reading mission was my confidence.

> If we want to improve student achievement in mathematics and shift mindsets and beliefs about what it means to be "good at math," we need to step up our game.

When I challenged students to read a new genre, I ordered a second copy of the book and read along with them. Each morning, I asked if they had gotten up to a certain part in the story yet or checked to see how they felt about a character's decision. When a student wanted to read a book that was a bit of a stretch, I offered the graphic novel first so they could become acquainted with characters and the storyline. When assigning students to book groups, I considered their likes and dislikes, personalities, reading strengths and stretches, and, of course, read all of the books in advance. When too many students were stumbling with deciphering new words, I signed up for Orton-Gillingham Training and introduced our etymology notebooks so we could incorporate word study in a way that would support all students. Because reading is essential to student success, I would go the distance to prepare my students to become competent readers. And because I loved reading and was confident that I could convince others to love it, I successfully converted many into confident lovers of reading.

Once I became more competent with math content and pedagogy, my confidence became a guiding light in teaching my students math as well. Students and colleagues were inspired by my boldness when letting them know that they will learn to love math at best and not be intimidated by math at least. Gone were the days of presenting mathematics as a list of rules and steps to memorize or a page in a textbook. Mathematics made sense to me, and I wanted that for everyone I came in contact with or had the pleasure of teaching. It was important to me that my students have positive and playful experiences with math.

Dry lessons from the curriculum had met their match because I was determined to win my students over to the math side and confident that it could be done. It was a LOT of work at first, but in time my passion for mathematics led to the same level of enthusiasm that was previously observed everywhere but math class, to become a regular part of the math experience. Competence and confidence worked hand in hand, as I strove to transform math teaching and learning for my students and colleagues. Soon, my students laughed in math class, talked about math all the time, and, most importantly, understood math concepts deeply. Before long, others in the school community wanted in on the Mather Movement!

Building Community Through Collaboration: Shifting the Culture

No one wants to collaborate when they are feeling insecure about what they know or don't know. As elementary educators, some of us chose to teach younger children to avoid "real math," and so we don't feel safe collaborating with math-loving folks who don't have compassion for people who don't have a solid math foundation. Young children have fragile math identities, and so do many adults, so let's not forget that building a community of mathers requires us to meet everyone where they are. Whether you have always loved math and continue to sharpen your skills or you are on the brink of developing a new relationship with mathematics, collaboration will enrich your experiences if done well and with purpose.

> **TIP**
>
> To encourage collaboration, have students work on a problem independently before they share with their group. Establish a norm that every student must contribute to the poster, sharing their ideas, and be able to explain any part of the representation and mathematical ideas. Then listen to them explain their strategies to one another and make connections in preparation to share with the class.

Mathnote
Learn to Chill-LAX

In lacrosse, there isn't time to chill on the field, but the game is often referred to as LAX, which is slang for lacrosse and translates to "the stick." Lacrosse is a fast-paced sport that is not for the faint of heart. One must be agile and fit to run the ball and catch, carry, and shoot it into the goal. To score, one must attend to the angle at which you shoot and to keep the ball in the net, Newton's laws of motion are in full effect. This popular sport began in the early 12th century with Indigenous tribes in Canada and North America and was originally called stickball. Lacrosse was played as a means for settling disputes, training for war, or for recreation and the pleasure of the Creator.

> **TIP**
>
> Have simple math challenges with your colleagues each week that students can support their teachers with solving. These can be posted on a bulletin board outside your classroom with a bag or folder for submissions. At the end of the week, announce all who "won" the challenge.

When adopting a new curriculum, there is this moment of "We're all in this together!" that is worth capitalizing on to build a strong math community. Of course, some will be more comfortable than others, and levels of competence and confidence will vary. But an unfamiliar program invites collaboration. Even when we are not adopting a new curriculum, we can find common ground to work together to create a sense of belonging and establish a positive math culture. Our actions and decisions will signal students that we believe wholeheartedly that we are all mathers.

It goes without saying that planning together can be impactful for teaching teams, but collaboration is also noticeable to the students. The more our teachers worked together to plan math lessons, the more we popped into one another's classrooms. The more comfortable we became, the more often we joined forces and shared resources. It became a normal occurrence to wrap up a unit with a scavenger hunt or math stations that were set up across classrooms. Collaborating to create these experiences meant that we shared the load with planning and set up, but it also meant that our students were exposed to different teaching styles and had opportunities to mingle with students from other classes. Mathematics was pushing us to collaborate beyond our class communities, and we were loving it.

ACTIVITY TO TRY
Game Stations

Choose a unit to gamify or create stations for with a colleague for the end-of-unit review. Create several stations to set up in each classroom, doubling the number of stations students get to experience. Mix the students across the two classes to travel from station to station together in small groups. Provide students with a clipboard and a BINGO type board to keep track of stations completed. Some stations should be directly connected to the concepts in the unit, some should be for review of concepts, some should focus on fluency, and some should be fan favorites.

Use a timer to ensure that each group remains at their station for the same length of time, and rotations occur simultaneously. Let students know they can play multiple rounds if time permits, and be clear about your expectations for clean-up. Be sure students are introduced to the types of activities that will be included in advance, so there isn't a need to have detailed explanations. Instead, have directions at each station with visual supports or prerecorded audio with directions students can listen to at the start of each station. This might need to happen across two days, but it could also be done at the midpoint of the unit and continued at the end of the unit. It is a lot of work the first time, but the key is to create stations for use with your students all year. This way, when you collaborate with a colleague, their activities are fresh for your students and vice versa. And, after these special joint station days, you can continue to use them in your classroom as learning stations for ongoing practice and review.

Math Without Borders: We Are All Mathers!

Extending the math community beyond the classroom is essential to shifting the math culture in a school. Any and all adults who interact with students need to encourage a growth mindset in mathematics and lead by example. Students need to see everyone getting excited about mathematics, or at least invested in their success. At the end of each year, our review involved all of the core concepts and required a coordinated effort across the campus. There were math riddles in the classroom, math posters with problems to be solved in the hallways, Jenga fraction towers in the library, multiplication squares on the blacktop, gumball estimation jars, geometry math art stations, and more.

The first year I set it up, there were many sideways glances, but in time I gained the support of teachers who offered to staff stations, lend their empty labs, or help tally up scores. My year-end scavenger hunt became a fan favorite that even our principal looked forward to each year. We had a fabulous time, and students looked forward to the math mysteries ahead with confidence. Older students who had experienced it in years past smiled as my current students scurried by on their way to solve the next challenge. When I left the classroom to write curriculum full-time, I returned to the school to set up the scavenger hunt for my colleagues and ran it so they could just enjoy it alongside their students. There is nothing like witnessing Pure Math Joy!

As we continue to push boundaries and spread math love, hosting Math Nights for families is another way to expand the math community. This is a wonderful way to gain the support of families and help them to see why this "new math" isn't so scary. I had the privilege of being a part of a Math Magical Fun Fest that was a districtwide event held on a Saturday. Parents, guardians, grandparents, and siblings spent a few hours playing games at stations all throughout the school building after an opening keynote focused on spreading math joy. There were pictures and prizes, and I was thrilled to pop into the library as the superintendent read math stories with a group of students and their families. If we truly want to shift the culture in our schools with lasting impact, we need to include families.

 **ACTIVITY TO TRY**

Host a Community Math Festival!

In a community event with families and their students, we had the opportunity to engage with math tasks together, learned new ways of thinking about problem-solving, and laughed as children and adults estimated how many ducks were in a jar using a Steve Wyborney Estimation Clipboard activity. Everyone participated willingly, as they had multiple chances to get closer with their estimations each time I asked, "How many ducks?" As we wrapped up, we used call and response to boldly chant, "I am a reader. I read for fun and for information. I am a writer. I write stories with my imagination. I am a mather. I use math to make sense of the world." Grandparents, parents, guardians, friends, teachers, and children joined in chanting about being competent readers, writers, and mathers. By the time the chant was introduced, they believed every word and stood tall as they added mather to their vocabulary and embraced mathing as a part of their identities. In each classroom and in the hallways, there were stations set up with math games and puzzles to solve courtesy of the Julia Robinson Mathematics Festival (Robinson, n.d.), which is an organization that helps schools host math festivals, family nights, and other math events designed to inspire math joy.

Source: istock.com/nd3000

Teachers hold the power to disrupt the narratives that play on repeat with mathematics as the villain in the story. The math vibe is set by us, and if the math lovers are outnumbered it can be an uphill battle.

As you develop your comfort, grow in competence and confidence, and begin to see changes in your classrooms, plan for the resistance that may come from within. Sometimes others will decide, "You're doing too much," when in reality they are wondering how you did it. Keep in mind that resistance is sometimes the result of fear, so we need to practice patience and lead with humanity. Reflect on your own journey and offer stories that include your successes and failures along the way. Sometimes leading the way means pulling back the curtain so others know they are not alone. The vision is for all students to be competent readers, writers, and mathers, so we need all teachers to join the Mather Movement and become the mathers our students need us to be.

> Sometimes leading the way means pulling back the curtain so others know they are not alone.

WHERE'S THE MATH IN THAT?
Coaches and Trainers Are Mathers!

Source: istock.com/bmcent1

Angelina Perrone is an energetic trainer at the gym where I work out. But one day, she made the mistake of saying she doesn't really do math, only "Girl Math." Let's just say, she was not ready for the response that came

(Continued)

(Continued)

after that. In the visits that followed, Angelina greeted me each time with a smile, and we joked about all the math that would be happening in boot camp that day. A few weeks in, she was hooked and even agreed to be interviewed for this book.

In this interview with Angelina, scholar, athlete, coach, trainer, and master of business administration (MBA) holder, we explore the connections between athletics, fitness, and mathematics. Angelina shares how she never considered all the math she was doing every day, but now after meeting the Queen Mather, she can't stop thinking about it and sharing it with anyone who will listen.

Deborah: Tell us about your journey to becoming a trainer and a coach.

Angelina: I grew up playing soccer and lacrosse. My mom had me try lots of things, like dance, field hockey, and basketball. I basically played a sport in all four seasons all the way through high school, but decided to narrow it down to two when I went to college. I played field hockey and lacrosse in college, which wasn't easy at a Division 2 school. Even though I excelled at both, my love for lacrosse never waned, so I knew I could never let it go.

I didn't know if I wanted to be a coach when I was in college because I was more interested in marketing, social media, and project management. Right out of college, I got into sales with a sports publications company, but quickly realized it was not for me. At first, the job was interesting, but cold-calling was not something I could see myself doing long term.

I decided that I would start looking for coaching positions. I had the opportunity to be an assistant coach at a college on a volunteer basis with hopes of going full-time, but that didn't happen because of their budget. That's when I started thinking about going back to school. When I found a position as a graduate assistant, I was able to be a student and a part-time coach which was the perfect combination for me at that time. That's how I got my MBA and landed a job with Catawba as their lacrosse coach.

My trainer job at Burn was unexpected. I was training there and was approached about joining the fitness trainer program. At first, I resisted, but they encouraged me to try because they thought I would be a good fit. I studied and trained and got the job. It was a whirlwind, but now, I am a college lacrosse coach and a trainer at Burn Boot Camp.

Deborah: What type of student were you growing up?

Angelina: I was very much one of those kids who wanted to learn and do everything. I wanted to have fun with my friends but also stay on top of my work, even at a young age. I needed to be organized to do well and keep up with everything, so I developed that skill even in elementary school. My downfall was that I always wanted to be the center of attention. It was important to me to be the top student and the class helper.

Deborah: What was your relationship like with math growing up?

Angelina: I was never good at math, but I think I blocked it out of my brain in elementary school. As I got older, I became more aware and remember chatting with friends about how none of us understood what was happening in math class. We were frustrated, so we decided to work together to prepare for a test but we got in trouble when the teacher found out. She made us retake it, but we were so lost and confused that studying and retaking the test didn't make it any better.

Even when my teacher explained a problem to me, I would think, "I got it," but as soon as she walked away, I realized that I still didn't understand. It began to feel like my teacher was always frustrated with me. I couldn't communicate what I needed, so my teachers felt like they couldn't support me. I just remember feeling like a failure in math.

Deborah: That had to be so frustrating. How do you think your relationship with math influenced your decisions about your future?

Angelina: It definitely did impact what I wanted to do, but not where I went to college. I knew I was not interested in seeing a lot of math in whatever I decided. I had no plans to minor in math or choose electives that had anything to do with numbers. Number talk was like a foreign language to me. When my friends who were accounting majors were chatting about their classes, I had no idea what they were talking about. Basically, whenever possible, I avoided numbers and only did math when absolutely necessary. Of course, later when choosing to pursue a business degree, I knew some math would be involved. The only math I was concerned with when choosing my college was which school was giving me the best athletic scholarship, so I wouldn't have to spend as much on my education.

Deborah: It is ironic and hilarious to me that you wanted to major in business AND avoid math. Even as a trainer and coach, I think it's interesting that you believe you could successfully avoid math. So, now that you are all grown up, what connections

(Continued)

(Continued)

do you see between your roles as a trainer and coach and mathematics?

Angelina: Well, as a coach a lot of the math comes in with the administrative duties. Funny enough, my head coach makes me do the math for our budget. On the field, math comes up all the time. When we are doing drills or making teams for practice math comes into play. It's not crazy math like 4,000 times 235, but little math things like if we have a 3 v 2, man up situation, or a 4 v 3 situation, or a 77.

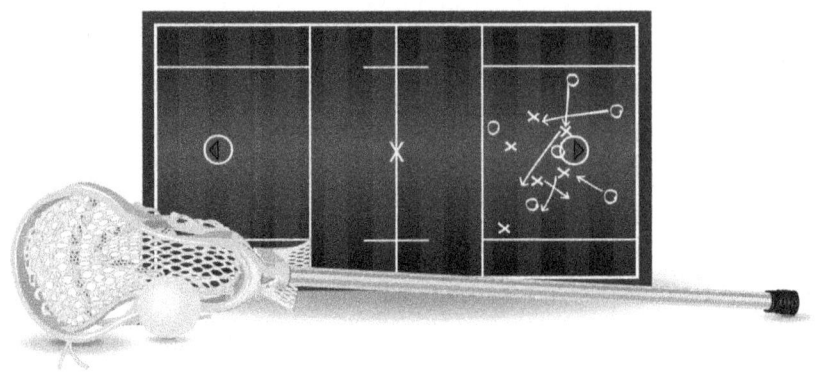

Source: istock.com/KeithBishop

Deborah: Please explain what a "man up" situation is and what these number combinations represent.

Angelina: In lacrosse, when we say a 4 v 3, it means there are 4 attackers and 3 defenders, so that's a "man up" situation. There are four quarters in a game, about 15 minutes per quarter, about 10 minutes for half-time, and 5 minutes between each quarter. I guess that's a lot of math right there. So, in a game when the clock is going down, I look at the scoreboard to figure out how to give my goalie some shots with at least 3 minutes left before half-time. If a player gets carded, green cards are 1 minute and yellow cards are 2 minutes, I need to determine on the spot when she can get back in the game and what that means for the team. Even though it's simple math, it's happening very fast. Even though people don't think about lacrosse as having lots of math, I guess it is very mathematical.

Even at the gym, I am keeping track of the last 40 seconds in a set, counting up and counting down depending on if I am counting someone's reps or counting time left on the clock, all at the same time. You have to tell your brain, "You can do two

	types of math at once. You got this!" So, it's more complicated math, even though it's kind of easy because the math is intertwined and happening in a fast-paced environment.
Deborah:	Well, it sounds like "Trainer Math" is its own category of math. It definitely requires a manipulation of time and space because you all count down, "Last, 10, 9, 8, get lower, 7, 6, 5, don't quit now, 4, 3, 2, aaaaaand, 1." You tell us it's the last 10 seconds, but we know it really isn't, but we go along with it because it doesn't seem unreasonable. Great trainers don't say 1 more minute and make you hold a squat for 5 minutes because in our bodies we can feel the difference and we stop trusting the count. We might even give up sooner. Counting backward is something we should definitely do more of in elementary school.
	So now you are on the other side, coaching student athletes. What would you tell a student who believes they don't need math to be a successful athlete?
Angelina:	I would laugh and then I would say, Oh, you do. You don't realize how much you are using simple math in everyday life. We might not walk around doing PEMDAS every second, but the simple math comes up all the time. Learning to do complex math makes you feel like the simple math is really doing nothing and you can rattle it off when you need it. Even if you don't fully understand all the math you learn, it's worth it to have some idea about how it works because you are going to see it again. Sorry to break it to you, no matter what you choose, math will be a part of it. Welcome to life. As a professional athlete, you need to at least know how to negotiate your contracts and keep track of your money so people don't take advantage of you.
Deborah:	Very true! You coach girls, and I know you've shared that many of them get a kick out of the "Girl Math" trend. You've also shared that you even found it funny, at first. But as a coach and role model, how would you encourage a young lady who really believes she can't be good at math?
Angelina:	With my players, we have been doing a lot of mindset work based on the book *How to be a Coffee Bean* by Jon Gordon and Damon West which focuses on creating a positive environment. So, they are familiar with the concept of shifting to a positive mindset. I would make sure they know that I believe in them, and that even though math can be hard, you have to keep trying. It would be easy to connect it back to the mindset work we are doing to be better lacrosse players. I would remind her that the drive and determination it takes to get better at a

(Continued)

(Continued)

                    sport can be applied to everything in life we want to achieve, even math. We are all mathers here!

Deborah: I love that! What would you tell your younger self to convince her she is a mather?

Angelina: I would tell my younger self that the same attitude you have on the field to push through failure and to work hard to be the best, is the same attitude you need to have to get better at math. Failure in the classroom is viewed as such a bad thing, but it doesn't have to be. Getting a B, C, or D on a test is not the end of the world. When you fail, use your resources to get better. It's okay to fall, but you have to get back up and keep pushing. This is also what I tell my players and it's how I encourage people in the gym. It's when you are uncomfortable that you are growing and improving, so get comfortable with being uncomfortable.

When students demonstrate the ability to persist through challenges in any area, we can remind them that math is no different. As we disrupt the myth of the math brain and help students realize that mathing is something we are all capable of, they can change their mindsets and beliefs. So many scholar-athletes would never accept defeat on the court, in the pool, or on the field, so why should we let them accept a math failure as their destiny. Anything worth being great at is worth fighting for. Since mathematics is needed to function in society and is connected to any dream we can dream, let's help our students see themselves as mathers and encourage them to never give up.

TIME TO REFLECT AND TAKE ACTION

6

"Do as I say, not as I do," is something my second-grade teacher often said to us. She was very strict. She cautioned us that drinking Tab cola everyday was bad for you. Yet, she had Tab several times a day while we were watching. It was a requirement for us to be kind and patient with one another, but she became frustrated easily and used a coarse tone to correct students regularly. Mrs. J demanded respect, but what she actually got was students who were fearful of consequences. This meant if she wasn't around, we could do whatever we wanted. Substitute teachers were not fans because of the mismatch. If Mrs. J is so strict, why are her students so "wild"? In all honesty, our class wasn't wild, just happy to breathe freely and chat with our friends. I guess in the era of "Children should be seen and not heard," too many giggles equated to being out of control.

If we want children to embrace mather as a part of their identities, we MUST be mathers ourselves. We cannot say, "You can do it!" and complain to our colleagues, "This math curriculum is bonkers. I don't even understand the problems." While it may be true, we need to model a growth mindset by admitting when math is confusing or challenging and acknowledging that we may need to think about how to solve a problem and come back to it later. Or maybe, admit that you will ask Ms. Jackson down the hall how she approached the task. By honing our math skills and planning our lessons with purpose, we nurture our own math identity development. Children will do what we say, but they are also watching what we do. Let's invite them with our words and our actions to be mathers.

1. What is one small step you can take to develop your comfort and confidence with teaching mathematics?

2. Which math routine will you share as an opener at your next team meeting?

3. How does your language impact your students' view of mathematics? What will you change?

4. Which math concept is your least favorite to teach? How will you make it more interesting and engaging for your students (and yourself)?

5. Who can you invite to play around with math problems with you? (Get those manipulatives ready.)

 MATHFIRMATION

Sometimes all we need is a little reassurance that we are not alone and we suddenly find the strength to move on. We ask students to repeat after us for pledges and practice, and they say words they don't always believe to be in compliance. The cool thing is that if we say something over and over again, eventually we start to believe it.

At the start of math class or during the morning meeting at the carpet, call in your students by asking with confidence, "Where are all my mathers?!" Teach students to reply, "We're right here!" Then launch into a short Mathfirmation:

Teacher: Math is for

Students: Everyone

Teacher: Sometimes it's hard

Students: But i

Teacher: Math

Students: And

Teacher: Math

Students: And

Teacher: Are y
           math

Students: Yes,

CREATING THE CONDITIONS FOR POSITIVE MATH (LEARNING) EXPERIENCES

Source: istock.com/FatCamera

Elementary school is a place where students learn to be a part of a community that is like extended family. Children spend most of their waking hours with this group of people called "their class." It takes time to adjust to the norms of this community, where others' opinions matter, and hurt feelings happen even when we didn't mean it. There are consequences for our actions that may not align to what we experience outside of school. Needless to say, being an elementary student is an adjustment. It takes time to settle in at the start of each school year and after any extended holiday break. Everywhere they turn, children are learning a new set of rules based on the adult who is in charge. Sometimes it's easy to shift from one environment to the next, but sometimes the rules and expectations are in direct opposition of one another. So, what can we do to support young children with joining our communities? Get rid of the rules!

> I know what you're thinking. Chaos will ensue and no learning will happen.
>
> It won't. I promise.

At the start of each year, we have a new group of students with varied personalities, attitudes, strengths, and needs. It is not possible that one set of rules will be exactly what is needed for every group, every year. Early in my career, I did have class rules, and for the most part students obeyed. But the more I thought about school as the place where children spend most of their time when they are awake, the more I considered the importance of my students having a say in our classroom environment. After a little soul searching and a little research, we adopted a Village Mentality.

Many are familiar with the saying, "It takes a village to raise a child." This implies that the children of the village belong to all of the adults, and everyone is committed to looking out for the village children. In several cultures around the world, being a village is so much more. In a village, there is often a philosophy that "we comes before me," signaling village members that they should consider what is best for everyone not just what they want for themselves. In South Africa, there is a common phrase to describe the interconnectedness of a community. *Ubuntu*, which comes from the Zulu and Xhosa languages, means, "I am, because you are." It encapsulates the ideas of compassion, kindness, caring, and connectedness that are needed to build true community. It implies that the village is the sum of its people and together its members struggle or thrive. Everyone in the village makes a difference, so they consider others before making decisions.

Source: istock.com/Ridofranz

In an effort to lead with *Ubuntu*, I set up the classroom to be inviting and intentionally incomplete. Before students arrived, I didn't cover every wall and decorate every corner because I wanted them to know that I was waiting for them. On the wall were the words, "It takes a village . . . " At the beginning of the year, we took time to discuss what this means. Students recognized that what I meant was that we all have a place here, and everyone's actions impact the village. It meant that together, we are smarter and can solve problems when we leverage everyone's strengths. It meant that every member has a responsibility to consider how their behavior, good or questionable, impacts others. It meant that everyone has a voice and that when one fails, we all fail, but when one wins, we all win. It meant that I was the adult, but I wasn't the only one making decisions about how our village should function. It meant that it wasn't up to me to establish class rules, it was up to us to create our Village Agreements.

The process of creating the Village Agreements took time, but it was time well spent. Children generated a list of five things they need to feel safe and focus on learning. They did this individually, and then discussed their lists with their table groups of four students. Together they compiled a list of five things for their group. They reasoned with one another, explaining why the need they included mattered and negotiated which would make the group list. In most cases, there was a lot

> Before students arrived, I didn't cover every wall and decorate every corner because I wanted them to know that I was waiting for them.

of overlap and students did a wonderful job combining or rewording statements to ensure everyone's ideas were included. The only way a need could be left off the list was if the original contributor agreed that it was addressed by another statement or decided it didn't need to be in writing. The next part of the process was for me to serve as scribe, as groups shared their lists. The goal was to have no more than seven agreements. Students engaged in the same negotiations and revisions they did with their small groups with the whole class until there was everyone could live with.

Mathnote
Math by Design

Diarra Bousso Gueye is a mathematician, math teacher, and fashion designer from Senegal. This brilliantly creative mathematician went from working on Wall Street to designing clothing using math concepts and equations to create bold prints. Diarrablu is the name of her brand, which features designs like the Joal print that was inspired by exponential and quadratic functions. Gueye proves through her designs that mathematics is meant to be a part of the creative process and that the fashion industry can employ sustainable practices.

Fashion Inspired by Math: Diarrablu

https://qrs.ly/13gnmtq

The final step took place the next morning, when I returned with the Village Agreements printed on scroll paper. Every student received a copy, and there was a laminated copy to hang on our community board. Everyone stood and recited, "We agree that . . . " as we read each of the agreements, and then each person signed their copy and placed it in the front clear pocket in their binders. These young children started the year feeling respected and they created agreements they could live by, a first experience with collective agency. This means that all year long they held one another accountable by reminding a friend what they had agreed to when someone was out of line. It was beautiful to watch the negotiations and listen to why students felt strongly about having it in writing that "We agree to listen when others are speaking" or "We agree that putting away the materials you use when you are done helps keep us safe." Each agreement was stated from the positive and every student knew they had contributed to the agreements that would support a safe learning environment.

Source: istock.com/FG Trade

The agreements set the tone for the school year, but regular Village Meetings kept us focused on our collective goals. Every Monday morning, we met to discuss how we were feeling, to share goals for the week, to share good news, or make requests. Every student created a Mood Meter based on the work of Marc Brackett, founding director for the Yale Center for Emotional Intelligence, and kept it in their binders to help them consider the precise language they could use to describe their emotions. Instead of saying they were *tired*, students tried new vocabulary like *drained* and explained how the events of the weekend led to feeling this way. The students learned so much about one another, and it helped them grow in compassion and empathy. They also supported one another with academics with patience and care, especially when someone had shared about feeling glum. It's also no coincidence that these experiences influenced their writing and enhanced their understanding of character development in the books they were reading, while preparing them to be assets to the class community.

> **TIP**
>
> Use the How We Feel app or other resources that have tools to help students attach precise language to their feelings. Students didn't have the app. I had it on my phone, so they came to me if they were looking for synonyms for an emotion.

Someone once asked me how to build a math community, but they shared they knew how to build a community, just not a "math" community specifically. My answer was simple: "Teach students how to work together, laugh together, fail together, win together, and care for one another. Math is about making connections between strategies and mathematical ideas, but it is also about making connections with everyone around you." Solving problems is hard work, but doing it together makes a difference. The students in my class often reminded one another of our hashtag, #TogetherWeAreSmarter. In a nutshell, to build a math community, focus on building a community of compassionate people, so they can math together and learn from one another, and begin the journey toward global citizenship.

> *Teach students how to work together, laugh together, fail together, win together, and care for one another. Math is about making connections between strategies and mathematical ideas, but it is also about making connections with everyone around you.*

Elementary school can sometimes be a place where cliques create conflicts. Certain students gain status because of the clothes they wear, their height, or some other unspoken criteria that gives them social power. This can start as early as kindergarten, and if left unchecked can interfere with learning, especially mathematics. Since math is perceived to be easier for the "smartest" students, being the first to answer is often rewarded with social capital. Students begin to notice Rochelle always knows the answer right away, so she "must be the smartest of us all." But what if the Village Mentality establishes a norm that faster isn't smarter, and it's not impressive to shoot your hand up first? What if we understand that thinking deeply is expected and understanding new concepts takes time, so it's okay not to be first?

In our village, we talked about how we process information at different speeds, and that it doesn't mean someone is not as smart because of a slower processing speed. We also learned about how some mathematicians thought deeply about problems for hours, days, or years, contemplating approaches and possible solutions. They agreed that it would be disrespectful to interrupt someone's thinking by making noises and waving your hand in their faces. This was connected to our Village Agreements, so everyone worked hard to honor the think time of others even when excited to share. When a prompt was given, students knew either I would offer quiet think time and then hands could go up or I expected quiet thumbs up at your chest where it wouldn't be distracting to others. This culture of mutual respect spilled over into other areas, and students became more patient when others needed more time. This became the new norm.

It takes time to break the habit of competing to be the first, so I had to work on my reaction to hands shooting up. I also had to take answers from several students, even if the first one was correct. Students learned to use hand signals to show whether they agreed or disagreed with a response. All solutions were valued, and together we determined which solutions were correct but also why. Students often revised their responses after hearing the reasoning of another student, not because they assumed their answer was incorrect since they were the only one who believed it. The students learned the value of feedback and often received it better from their peers than from me. I loved that and used it to my advantage. During a math block when students were solving story problems, Manny decided to try a new strategy to find the difference between 5,000 and 2,347. Instead of regrouping, he subtracted 1 from 5,000 to make 4,999. His final answer was 2,652. When I asked if there was anything he needed to do about the one he had taken away, his response was, "Nope." I knew the village would lovingly correct this partial conception, so I asked Manny to share during our lesson wrap-up. Manny shared proudly, as his classmates nodded along. When he was done, hands went up. He chose someone who led with, "I like your strategy. It made it much easier to subtract," and then gently asked, "Don't you need to add back the 1 you took away

Source: istock.com/Zinkevych

TIP

Teach students silent signals for "I'm ready to share" or "I'm still thinking." We also used hand signals to communicate levels of understanding. We used thumbs up if they understood and were ready to work independently, thumbs sideways if they still had questions, and thumbs down if they wanted a little more practice or support before getting started.

since you took away 1 too many?" Manny paused, and exclaimed, "Oh, yeah! I see. It's 2,653." Other students gave the signal that they agreed and gave a few high fives to Manny. I would find out at the end of the period how he felt about this experience.

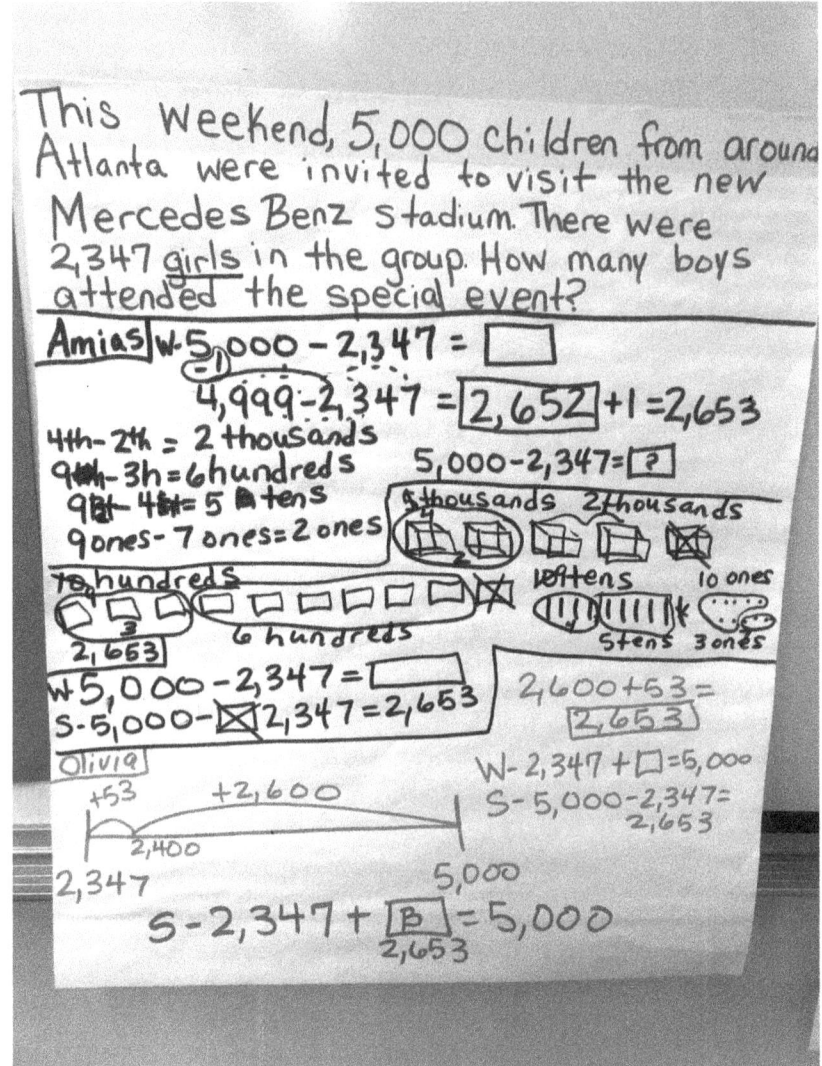

Everyone had a math journal that was used for writing reflections at the end of class. I enjoyed reading about the new strategy they were planning to try or how a student noticed something new when a classmate shared their thinking. One of my favorites was reading the reflection that Manny wrote that day. "Today, I got to share during math circle time. I was proud to share my strategy, but my answer was wrong. Next time I will remember that I need to put back the one. I cannot wait to have a chance to share again." If I had just told him to change it, I don't believe there would have been the same level of understanding. Manny wasn't defeated; he was inspired. That was a result of the beautiful community that had been built and the Village Mentality that we are smarter together.

Source: istock.com/Ridofranz

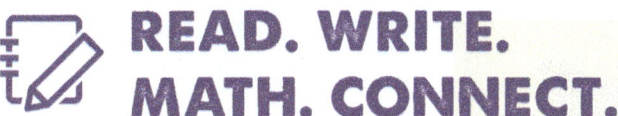

READ. WRITE. MATH. CONNECT.

Students are encouraged to write in journals during the literacy block. This reflective practice can also be powerful during math class. Let's help students see the value in taking time to pause, reflect, and plan for future changes to improve in mathematics. Words do belong in math class, and you can never have too much practice with writing. Mathers write and Writers math!

The intentional work early in the year to establish a sense of belonging and safety, supported students with making choices during interactions in math class. Students learned how to respectfully disagree, critique the reasoning of others, justify their own reasoning, and accept when they were wrong without falling apart. All students contributed to math discussions because they recognized that it takes a village to solve problems.

During core math instruction there isn't always time for lessons in math community etiquette, but warm-ups are the perfect time to build community and invite all voices into the space. Even students with solid academic identities may have a fragile math identity that could use a little nurturing. The launch of the math block is a great place to do it. Depending on the curriculum, there may be time built in at the beginning of each lesson for a warm-up or launch, but even if there isn't, this invitation plays a vital role in building confidence for students so we should make time for it. Whether the activity is directly connected to the lesson focus or it is an opportunity to practice basic math skills or fluency, carving out 10 minutes each day to engage students in a low stakes task and prime them for learning is a necessary investment.

 **ACTIVITY TO TRY**
Problem of the Day

At least once per week, pose a "Problem of the Day" that students can think about and solve. Give additional information, hints, or clues throughout the day, but don't give it away. Have students turn in responses in a basket or folder so you can check during the day. Before the day's end, share the great thinking you witnessed and the final solution.

As we continue to dispel the myth of the math brain, it will be necessary to remind students that math is for everyone. A daily reminder is just the thing to bring students back from a failure from the day before, confusion over homework, or the ongoing battle with stereotypes. Starting math class with warm-ups and engaging in group tasks regularly is a constant reminder that we believe all students belong in the math circle. Whether figurative or real, a circle captures the equity we desire in math class. All students, points on the plane, are equidistant from the center. Everyone has access to the central learning if we set them up for success and help them build bridges to connect their understanding. And as part of the circle, the teacher is a facilitator of learning not the focal point, not the knower of all things. We have to be intentional with relaying the message that we are all mathers and have a place in the math circle.

In the math circle, all ideas are welcome and mathematics is about exploration and discovery. I am not advocating for students to ONLY "discover" math, but I am saying they don't need as much direct instruction as some would have us believe. Just as we have circle time in literacy, we can have math circle time to share math ideas and play circle games. A circle also lends itself to discussions and debates and can serve as the practice ground for listening, observing, analyzing, and verbalizing opinions just as we do in discussions in language arts. Math discourse deepens conceptual understanding, so we need to get the children talking.

When we talk about mathematical ideas and reason together about why something is incorrect, students learn that mistakes are opportunities for growth. If students are in the habit of sharing their developing ideas with a classmate, they learn that mathematical ideas can be revised based on new information. Before long, they will be making conjectures and proving them using nonexamples and counterexamples. When students hear different perspectives and solution paths explained, they recognize that even if there is one right answer, it doesn't mean there is one right way to find it. In horseracing, the winner's circle is reserved for the winning horse and jockey, so they can receive their awards and have their picture taken. I like to think of the math circle as a place

> **TIP**
>
> One way to reinforce the belief that we are all mathers is to give students the opportunity to create and/or lead math warm-ups. With a little coaching, every student can have a turn leading a math routine they have regularly participated in, or at least co-lead with a buddy. This role is great for the student leading and sends a message to the other students about who can be "good at math."

Whether figurative or real, a circle captures the equity we desire in math class. All students, points on the plane, are equidistant from the center.

> **TIP**
>
> Establish routines for bringing students together or for getting them to stop talking, freeze, and listen. If students were working collaboratively, they knew that when I said, "Hey," they responded with, "Ho," two claps, and "Shhh."

for mathers to come together and win together, breaking down the barriers of exclusion and welcoming all students in to receive their reward . . . #MathJoy!

Source: istock.com/FatCamera

READ. WRITE. MATH. CONNECT.

Sharing unfinished ideas is a regular practice in literacy. As you brainstorm ideas for the next story you will write or revise a story based on feedback, there is an expected grace that comes with exploring the possibilities you might pursue. Mathers deserve the same grace as they share incomplete solution paths, inventive strategies, and partial conceptions with classmates. We accomplish this in writing class, so let's work toward normalizing these practices in math class. Math joy happens when all ideas are honored and respected and students learn to receive feedback without fear of judgment.

Have you ever thought about how you could level the playing field when practicing math fact fluency or computation skills? What if I told you that making math a sensory experience helps? It also involves listening skills and focus, so missing an answer doesn't necessarily reveal a computation weakness; it could mean a student heard incorrectly or wasn't focused. It even works with adults! It is a favorite activity of mine and one that students enjoy. You can do it as a warm-up or because you have a few minutes to spare. Mathing through listening will always be a hit.

As I get ready to share the first problem, students sit up tall or put their heads down. Some close their eyes, while others take a few deep breaths. An observer might wonder, "What is about to happen?" I set up my stainless steel bowl and plastic tub and grab the base-10 blocks from the shelf. Students retrieve their whiteboards, a marker, and felt square to erase as we go. They're excited, but a hush comes over the room because they know that listening is a huge factor with this computation practice. I use the stainless steel bowl for the ones, so they make a *clink, clink* sound. For the tens, I use the plastic tub, so they make a *clunk, clunk* sound. For the practice round, students look up, as they get used to the sounds and prepare to calculate. They hear, *clink, clink, clink, clink* and *clunk, clunk*. When I instruct them to hold up their whiteboards, a quick scan of the room reveals that every student knew the answer was 24.

Source: istock.com/SolStock

Now, students look away or close their eyes so they rely on their listening and mental calculations for the first round. Of course, I start off with a certain number of ones, and then tens. Soon, I drop in more than nine ones, so there is an opportunity to regroup. With each round, it gets a little more challenging. I shorten the pauses in between dropping blocks and change the order, dropping ones, then tens, then ones again. As I scan the room, it is easy to see if a student is off by one (maybe a listening error) or off by

TIP

Felt squares make excellent erasers for student whiteboards. They are inexpensive, especially when purchased precut in bulk.

CHAPTER 7 • CREATING THE CONDITIONS FOR POSITIVE MATH EXPERIENCES   **183**

20 because of a computation mistake. Sometimes it might be both. It remains fun for the students because there isn't a moment of calling out who got it right or wrong. It's practice for them and information for me, so I can follow up later. Liljedahl (2021) encourages the use of nonpermanent writing spaces because students are free to try something, fail, erase, and try again. I found that my students grew more and more comfortable with making mistakes, even publicly, when they knew that in the next round they would have another opportunity to try again. They held up their whiteboards high with confidence that didn't wane when I announced the correct answer, knowing that it could've been a calculation error but it also could have been a listening error. Either way, there was no judgment and there were smiles all around.

ACTIVITY TO TRY
Mathing Through Listening

Mathing Through Listening is a great activity even in upper-elementary grades. Make sure you have containers made of different materials so the math tools make different sounds when they land. Start with just a few blocks (I use base-10 blocks), so students get used to keeping track and differentiating between the sounds. For younger students, you can use only ones, have them look while you drop the cubes, and,

> of course, slow down. Most students keep track mentally, but others will need to keep tally marks or some other form of tracking on their whiteboards. Feel free to adjust the rules as needed or later include hundreds.

Together!

Building community is about more than setting up the tables as groups instead of in rows. It requires us to consider all of the cultures, habits, lifestyles, and ways of being that will enter the classroom on day one. Will we be ready with books that provide windows and mirrors for **all** of our students? Will we choose **all** the right colors, posters, and games? Or is it possible that we are not the ones who should make all the decisions for creating the learning space? If we want all students to feel welcomed and experience a sense of belonging, what parts of the classroom set up can we intentionally leave undone so we can get our community members' input? Shouldn't we wait to see who shows up so we can seek out opportunities to include images that represent all of our learners? Shouldn't students be invited to bring in parts of who they are to add to our shared space?

> *Shouldn't students be invited to bring in parts of who they are to add to our shared space?*

We want all students to know we believe they can be competent readers, writers, and mathers, so let's create spaces that say, "All Learners Are Welcome!" without a poster. Maya Angelou said it best: "People will forget what you said, people will forget what you did, but people will never forget how you made them feel." Will our students **feel** welcomed every day when they are in our care? Let's make the effort to make sure they do.

>
> ## ACTIVITY TO TRY
> *Call and Response*
>
> For practice with following patterns and listening, do a call and response using claps to get students' attention. Teach students that at the end of a clap and respond, you will be sharing important information or giving instructions. When students are collaborating or moving around the room, clap out a pattern for them to clap back and wait for instructions.

WHERE'S THE MATH IN THAT?
Artists Are Mathers!

Naja Brooks has always been an artist. At the age of two, her fingerpainting of a horse actually looked like a horse! By the time she was three years old, her drawings of people included eyelashes and eyebrows, their bodies were symmetrical, and not only did they have 10 fingers and 10 toes, but they also had fingernails and toenails with polish. By the age of four, Naja was writing and illustrating her own books. She wrote stories using phonics-based spelling and included detailed images to provide the visual narrative. Crayons, colored pencils, paints, construction paper, canvas, or napkins—it didn't matter. This little artist could make a masterpiece with whatever was available. As her mom, I didn't realize how talented she was until her kindergarten teacher told me at our first conference that Naja was writing too slowly because every letter she wrote was its own work of art.

Over the years, there were many art workshops, summer programs, and apprenticeships. As a middle-school student, Naja asked when her school schedule could include more of the subjects she enjoyed. When she heard college would be the time when she could focus mostly on art, she started researching art schools. As a seventh-grade student, she proclaimed that she would be going to Savannah College of Art and Design for animation or illustration. And she did. It is my honor and privilege to include an interview with this talented illustrator—the illustrator of *this book*—not only because she is my daughter but because she is the creator of the Mather Kids, Kiana and Kareem, and one of the first members of the Mather Movement.

Deborah: What type of student were you growing up?

Naja: I was the type of student who tried my best in class, even when it felt like I was always falling behind. Growing up with undiagnosed learning disabilities meant that I needed extra help to stay on top of my work. Oftentimes, I had to clarify with educators how I understood what they were teaching me at the time because they didn't understand my process or my thinking. When tasks were too easy or completely unappealing, it would bore me. Luckily math was not one of those subjects.

Deborah: What was your relationship like with math?

Naja: I love, love, loved math! Multiplication problems were my absolute favorite. My classmates and I would come up with fun songs to remember different solving techniques and some of the games we played made math very engaging. Unfortunately, because of the stigma around math, unenthusiastic teachers

(Continued)

(Continued)

would paint negative pictures around the subject, especially if you weren't the fastest. By fifth grade, my speed was lagging, and my confidence died down. Fifth grade was the year when expectations around math and "your future" started to ramp up, and math was less about exploration or fun.

Deborah: How did you decide that you wanted to be an illustrator and animator?

Naja: I've always been drawn to the arts. I was always drawing and telling stories when I was younger; I danced and performed up until college. When it came to expanding my art skills, any and every medium was worth a try to me, and shaped where my art is today. Anytime we had an option to watch a movie or a show, my suggestion would always be something animated. I enjoyed analyzing the movie production and took inspiration from the art styles. The cartoons of the time were peak and they were incredibly inspirational. I thought my career path options were limited, but being an artist is the only thing I'd ever wanted to pursue in some way, shape, or form. What mattered most to me was that I wanted to be a visual storyteller who leaves a lasting impact on anyone who comes into contact with my art.

Deborah: What kind of art do you like to do?

Naja: My artwork has varied over time from trying all sorts of styles and techniques. When I am commissioned to do portraits as gifts, I enjoy stippling. However, the majority of my other work surrounds fantastical illustrations and animations. Fantasy is a passion of mine. Luckily, I've discovered that there is a wide spectrum of options to explore in illustration that I was unaware of when I was younger. I love the conceptual work that can go into one piece or a whole collection of works. This transfers well in advertising and graphic design. Some of the art that I am known for can fall into creating environments, character design, asset designs, and so much more. Because I've always been open to trying new techniques, I have a lot of artistic skills and talents at my disposal.

Deborah: What connections do you see between art and mathematics?

Naja: When it comes to art, there are the main principles that are foundational to creating an art piece: composition, perspective, depth of field, and the rule of threes. Ultimately, these rules and principles help guide an artist with whatever final artwork they're attempting to create. Even for abstract work, composition, measuring, and proportions are necessary to have a balanced work of art. For me, I've always loved breaking out the ruler and making pieces that required perspective. I relish in knowing what angles I'm using, and then pulling out to see the full piece come together. Zooming in on the intricate details and then zooming out to see how it all came together is always a fun and rewarding part of the process. In a nutshell, mathematics is an essential component to creating artwork, and being an artist means being a mather.

Deborah: What would you tell a student who believes they don't need math to be a successful artist?

Naja: I would tell them not to believe everything they hear. Just because some artists can get away with pretending math isn't an integral part of everything we do, including art, doesn't mean we should continue on that way. You don't have to be in love with a subject to understand how it plays a role in what you do. Unfortunately, even art schools don't take the math classes they offer as seriously as they should and require very little math. Mathematics only becomes a priority if it is what a student is looking for, and they make it a priority. However, I remember talking to illustrators who wished their schools had offered more perspective and composition classes, which have a heavy focus on mathematical aspects, because having that foundation makes a big difference in your craftsmanship.

Deborah: What would you tell your younger self to convince her she is a mather?

Naja: If I could go back and chat with my younger self, I would tell her that she and I already know we're mathers. I would caution her not to let anyone get in her head and weaken her confidence in pursuing or enjoying math. There is a magic to solving a problem, a joy manifested in figuring out how and why math fits into a lesson; don't let anyone kill that.

Deborah: I am so lucky to have you as my illustrator and my daughter.

Some students spend time doodling in class, and teachers fuss at them for being distracted. The truth is many of those students doodle so they CAN focus. A part of inviting students to bring their whole selves into our class communities means allowing artists to

(Continued)

(Continued)

show up with their visual talents. Naja was a star in art class because she lit up and came alive as she explored new art techniques and consistently created beautiful projects, but that was a special class that happened for less than an hour each week. She worked hard in all of her other classes and made good grades, but remained virtually invisible. I wonder how the math learning experience might change if instead of banning doodling, we highlight the strengths of these students and give them opportunities to support their classmates with visuals. Visual representations support conceptual understanding and help students showcase their mathematical ideas. We also know that artists are mathers, so let's help them make connections between art and mathematics early and give them opportunities to shine.

7 TIME TO REFLECT AND TAKE ACTION

As we consider building class community, we should view it holistically instead of treating mathematics as an outlier. Let's consider how to bring math into the fold as we build an inviting **learning** community with the intention of including all students. To build a community of readers, writers, and mathers where all students feel supported across content areas, we need to build a community of compassionate humans.

1. How will you invite students into the community on the first day of school AND grow our community together based on a Village Mentality all year long?

2. What mindsets or beliefs need to shift for you to build a math community that is more inviting and engaging for all students?

3. How can you adapt the warm and cozy reading spaces to also be inviting for mathing?

4. Whose voices do you anticipate will need amplifying as you create a culture of sharing and invite more math discourse?

5. What are you still pondering as you contemplate what a sense of belonging looks like in your space?

MATHFIRMATION

Setting intentions is a common practice that helps us focus on our goals. It can also motivate us to keep striving when things get challenging. I am of the mindset that every day is a new opportunity to try again, or improve on what we accomplished the day before. Each morning, we started our day with intention as we recited our Village Pledge.

> Today is a great day to learn something new.
> I am here to learn, how about you?
> I will work hard and play hard, be honest and kind.
> I will control my body and take charge of my mind.
> I will check in on my feelings, all throughout the day.
> And strive to be my best self in each and every way.
> Excellence is the standard, not the goal; yes it's true.
> Today, I will choose to be excellent in all that I do.
> Together we agree, here in the Village of 4-P that . . .
> This is going to be THE BEST DAY EVER!

(Continued)

(Continued)

We always closed
of My Life" by Am
best day of my lif
be the best day o
over, today has a
that every day is

8

SUMMING IT UP!

Source: istock.com/SDI Productions

You're almost at the end of this book. Are you convinced that students will thrive if their elementary school experience presents reading, writing, and mathing as the core academic skills? Or, are you still experiencing a bit of cognitive dissonance because this is contrary to everything you've ever been taught? The truth is, my goal was to plant a seed, not grow a tree, so even if you are still on the fence, I'm okay with it. Whether or not you immediately change your language, I am confident that you have a new perspective to consider that I hope will influence your view of mathematics. Math is not new, but we need to approach math teaching and learning in new ways. We must present math in ways that are more inclusive and attainable, so that all students know they have the right to be here and they are born to do math.

The Mather Movement is not about throwing out everything we know and believe about teaching mathematics. It's about writing the counternarrative about what it means to be good at math and creating a pathway to success for all students. Students who believe they can't do math are limited in their options as they choose careers, but even worse, they adopt a mindset that there is nothing they can do to improve their situation. The Mather Movement is a rallying cry to let everyone know they were born *mathers* and have everything they need to be successful. They can dream about a future in which mathematics will be a bridge and not a barrier. The sky is the limit when you believe you can do anything and put in the work.

> *Students who believe they can't do math are limited in their options as they choose careers, but even worse, they adopt a mindset that there is nothing they can do to improve their situation.*

I hope that as you read this book, you joined me on a journey of discovery. On this journey, I hope you were encouraged to **notice** without judgment the things that have shaped your thinking around mathematics. I hope that you were pushed to **wonder** what could happen if you let go of biases, misconceptions, and negative feelings you have connected to mathematics. Through reflection and discussion with your colleagues, I hope you took time to **think** deeply about what needs to change in order to make mathematics accessible and engaging for all students. I hope that as you gain a new perspective about what learning elementary mathematics could **feel** like, you are inspired to create a culture of curiosity and joy in math class for your students from this day forward.

Wouldn't You Like to Be a Mather Too?

I'm a mather. She's a mather. He's a mather. We're all mathers. Wouldn't you like to be a mather too? A *mather* is a person who uses mathematics to make sense of the world. The work begins with convincing yourself that this is true. Are you a mather? Even if math was hard for you and even if you don't like teaching it, will you work to shift your mindset? Or maybe, math was always easy for you, and you don't understand why we need to make such a big deal about it. Do you consider yourself to be a mather? Will you work to shift your beliefs about what it takes to be a great math student? Could you make a list of all the ways you use mathematics in your life OUTSIDE of the classroom? Well, let's start there! Whether you love math or have a disingenuous relationship with it, take inventory and determine what you need to do to be the best version of your mather self for students.

Teachers' math identities impact student math identity development, so there is a lot at stake. Math anxiety can be projected onto students by math-anxious teachers and math shame can be projected on students by impatient math teachers (Hadley & Dorward, 2011).

> A *mather* is a person who uses mathematics to make sense of the world. The work begins with convincing yourself that this is true.

TIP

Be honest about your relationship with mathematics and model a growth mindset as you take risks alongside your students. Stress the importance of working hard to get better and remind students that we all have what it takes to be competent mathers.

▶ What work will you do to ensure your students have a safe place to take risks and make mistakes?

- How will you invite all students into math conversations?
- How will you support students with making connections between mathematical ideas and their lives, and connections with one another?

Be the Change!

As elementary educators, we cannot let our feelings about reading and writing cloud our judgment when making instructional decisions. Reading and writing are not optional, so we must find ways to reach our students. They need to be competent readers and writers to thrive. There is so much support for improving reading outcomes, so let's leverage what we know and do to encourage reading and writing to improve mathing. We can shift mindsets and beliefs about mathematics if we help students see math in a similar light. As we plan to support students with developing core academic skills, let's add mathing to the equation and make a difference in the lives of our students.

You might be the only one in your building who reads this book, so what is the best first step you can take? It depends on where you are in your journey. My advice is to start with you. Assess what you need to be more comfortable with math. And once you have some level of comfort, start building community and breaking down barriers with math conversations and warm-ups. If it is a new practice, take your time. Don't try to introduce too many new routines at once. Seek out a colleague to take the leap with you and plan a few to try. Compare notes and encourage each other to keep going.

An easy adjustment to make is to add *mather* to your vocabulary. If we start calling children mathers from the beginning and are intentional with sharing positive math experiences, it will be etched into their identities. Mathing should feel natural. There doesn't need to be an appointed time to DO math. Help students discover all the ways math is happening in their everyday lives. They will begin to realize mathematics doesn't only exist in math class and hopefully they will start to see the creative side of mathematics. Celebrate all the different ways we math together from the simple things like sharing snacks, doing puzzles, cooking, or doing art. Use the "Where's the Math in That?" stories at the end of every chapter to share examples of mathing for fun and mathing for life.

Celebrate all the different ways we math together from the simple things like sharing snacks, doing puzzles, cooking, or doing art.

Source: istock.com/Chinnapong

 **ACTIVITIES TO TRY**

Get Everyone Involved!

Advocate for math community events to encourage a culture of curiosity and improve perceptions about mathematics.

- Petition your instructional leaders to bring in more mathematics professional learning opportunities to support teachers with building pedagogical content knowledge.
- Invite families in to share their math experiences and the ways they use math in their everyday lives or jobs to help break down barriers and dispel the math brain myth.
- Highlight mathematics across content areas and shine the light on cultural traditions that showcase mathematics.
- Rebrand mathematics in your community as something that requires hard work, but also something that can be the source of joy.

TIPS

Spread the word! As your students begin to enjoy math and grow in confidence with solving problems, don't keep it a secret.

- Share the work you are doing and invite others to visit your classroom.
- Display math projects, puzzles, and challenges on bulletin boards to spark interest and conversations.
- Recruit colleagues to join the Mather Movement.
- Host a math night to share some of the activities that have gone well or to solve math problems together. Find your people!

Hope for the Future!

Source: istock.com/gpointstudio

Imagine a world where children laugh in math class as much as they do at recess.

Imagine a world where children laugh in math class as much as they do at recess. Imagine a time when students plead for just a few more minutes to finish working on a math problem. Imagine a school where everyone believes they are mathers! What would it take to get students excited about mathematizing stories and writing their own? What would it mean for our society if all students were on track to be competent readers, writers, and mathers? As an elementary educator, what are you prepared to do to move toward a future in which it is common to hear children confidently saying, "Of course, I'm a mather! Isn't everyone?"

> I implore you one and all to join the Mather Movement with me.
>
> Stand up, stand out, and be the Mathers our students need us to be.
>
> So, what's it gonna be? What will you choose to do?
>
> I know I'm a mather! Wouldn't you like to be a mather too?!

Thank you for reading this book and going on this journey with me. Let's work together to disrupt the idea that math is optional. Let's stand together to write the counternarrative to the legacy of exclusion and gatekeeping. This work begins with our youngest learners, so this work begins with YOU.

Source: istock.com/SolStock

REFERENCES

Introduction

Dweck, C. S. (2016). *Mindset: The new psychology of success*. Ballantine Books.

National Center for Education Statistics. (2019). *Adult literacy in the United States*. Institute of Education Sciences. https://nces.ed.gov/pubs2019/2019179/index.asp

Chapter 1

Aguirre, J., Mayfield-Ingram, K., & Martin, D. B. (2013). *The impact of identity in K-8 mathematics learning and teaching: Rethinking equity-based practices*. The National Council of Teachers of Mathematics, Inc.

Boaler, J. (2022). *Limitless mind: Learn, lead, and live without barriers*. HarperOne.

Dweck, C. S. (2016). *Mindset: The new psychology of success*. Ballantine Books.

Gonzalez, L. (2023). *Bad at math?: Dismantling harmful beliefs that hinder equitable mathematics education*. Corwin.

Harris, P. W. (2025). *Developing mathematical reasoning: Avoiding the trap of algorithms*. Corwin.

Jansen, A. (2020). *Rough draft math: Revising to learn*. Stenhouse Publishers.

SanGiovanni, J., Katt, S. K., & Dykema, K. J. (2020). *Productive math struggle: A 6-point action plan for fostering perseverance*. Corwin.

Schunk, D., & Zimmerman, B. J. (Eds.). (2008). *Motivation and self-regulated learning*. Routledge. https://doi.org/10.4324/9780203831076

Sousa, D. A. (2015). *How the brain learns mathematics*. Corwin.

Starkey, P., & Cooper, R. G. (1980). Perception of numbers by human infants. *Science, 210*(4473), 1033-1035. https://doi.org/10.1126/science.7434014

Vakharia, V. (2024). *Math therapy: 5 steps to help your students overcome math trauma and build a better relationship with math*. Corwin.

Chapter 2

Champagne, Z., & Schoen, R. (2020). *Counting concepts*. Teaching Is Problem Solving. https://teachingisproblemsolving.org/counting-concepts

Duncan, G. J., Dowsett, C. J., Claessens, A., Magnuson, K., Huston, A. C., Klebanov, P., Pagani, L. S., Feinstein, L., Engel, M., Brooks-Gunn, J., Sexton, H., Duckworth, K., & Japel, C. (2007). School readiness and later achievement. *Developmental Psychology, 43*(6), 1428-1446. https://doi.org/10.1037/0012-1649.43.6.1428

Franke, M. L., Kazemi, E., & Turrou, A. C. (2018). *Choral counting & counting collections: Transforming the preK-5 math classroom*. Stenhouse.

Harris, P. W. (2025). *Developing mathematical reasoning: Avoiding the trap of algorithms*. Corwin.

Mononen, R., Aunio, P., Koponen, T., & Aro, M. (2014). A review of early numeracy interventions for children at risk in mathematics. *International Journal of Early Childhood Special Education, 6*(1), 25-54. https://doi.org/10.20489/intjecse.14355

Romano, E., Babchishin, L., Pagani, L. S., & Kohen, D. (2010). School readiness and later achievement: Replication and extension using a nationwide Canadian survey. *Developmental Psychology, 46*(5), 995-1007. https://doi.org/10.1037/a0018880

Wyborney, S. (2017). *Splat!* Steve Wyborney's Blog. https://stevewyborney.com/2017/02/splat/

Chapter 3

Carpenter, T. P., Fennema, E., Franke, M. L., Empson, S. B., & Levi, L. W. (2015). *Children's mathematics: Cognitively guided instruction* (2nd ed.) Heinemann.

Little Flower Yoga. (n.d.). *Home.* https://www.littlefloweryoga.com/

Luttenberger, S., Wimmer, S., & Paechter, M. (2018). Spotlight on math anxiety. *Psychology Research and Behavior Management, 11*, 311-322. https://doi.org/10.2147/prbm.s141421

Maloney, E. A., & Beilock, S. L. (2012). Math anxiety: Who has it, why it develops, and how to guard against it. *Trends in Cognitive Sciences, 16*(8), 404-406. https://doi.org/10.1016/j.tics.2012.06.008

Chapter 4

Bates, A. B., Latham, N., & Kim, J.-ah. (2011). Linking preservice teachers' mathematics self-efficacy and mathematics teaching efficacy to their mathematical performance. *School Science and Mathematics, 111*(7), 325-333. https://doi.org/10.1111/j.1949-8594.2011.00095.x

Bishop, R. S. (1990). Mirrors, Wwindows, and sliding glass doors. *Perspectives: Choosing and Using Books for the Classroom, 6*(3), ix-xi.

Boaler, J., & Williams, C. (2026). *Data minds: How today's teachers can prepare students for tomorrow's world.* Corwin.

Dictionary.com. (n.d.). Survey. In *Dictionary.com*. https://www.dictionary.com/browse/survey

Laib, J. (2024). *Slow reveal graphs.* Slow Reveal Graphs. https://slowrevealgraphs.com/

Maloney, E. A., & Beilock, S. L. (2012). Math anxiety: Who has it, why it develops, and how to guard against it. *Trends in Cognitive Sciences, 16*(8), 404-406. https://doi.org/10.1016/j.tics.2012.06.008

U. S. Department of Labor Women's Bureau. (2025). *Percentage of women workers in science, technology, engineering, and math (STEM).* US Department of Labor.

U. S. National Science Foundation. (2023). *Diversity and STEM: Women, minorities, and persons with disabilities.* https://nsf-gov-resources.nsf.gov/doc_library/nsf23315-report.pdf

Chapter 5

Artemenko, C., Wortha, S. M., Dresler, T., Frey, M., Barrocas, R., Nuerk, H. C., & Moeller, K. (2022). Finger-based numerical training increases sensorimotor activation for arithmetic in children—An fNIRS study. *Brain Sci., 12*(5), 637. https://doi.org/10.3390/brainsci12050637

Frey, M., Gashaj, V., Nuerk, H. C., & Moeller, K. (2024). You can count on your fingers: Finger-based intervention improves first-graders' arithmetic learning. *J Exp Child Psychol, 244.* https://doi.org/10.1016/j.jecp.2024.105934

Gottstein, G. (1999). *Perplexors: Mindware's best logic problems. Basic level.* Mindware.

Wright, R. J., Stanger, G., Stafford, A. K., & Martland, J. (2015). *Teaching number in the classroom with 4-8 year olds.* SAGE.

Chapter 6

Aguirre, J., Mayfield-Ingram, K., & Martin, D. B. (2013). *The impact of identity in K-8 mathematics: Rethinking equity-based practices.* National Council of Teachers of Mathematics.

Harris, P. (2025). *Developing mathematical reasoning: Avoiding the trap of algorithms.* Corwin.

Robinson, J. (n.d.). *Julia Robinson Mathematics Festival.* https://jrmf.org/

Wyborney, S. (2022). *170 new esti-mysteries.* Steve Wyborney's Blog. https://stevewyborney.com/2022/10/170-new-esti-mysteries/

Chapter 7

Liljedahl, P. (2021). *Building thinking classrooms in mathematics: 14 teaching practices for enhancing learning, grades K-12.* Corwin.

Chapter 8

Hadley, K. M., & Dorward, J. (2011). Investigating the relationship between elementary teacher mathematics anxiety, mathematics instructional practices, and student mathematics achievement. *Journal of Curriculum and Instruction, 5*(2).

INDEX

Able Math, 15
Academic identities, 11-12, 15, 17, 45, 47
Activities, 9, 18, 40-41, 52, 76, 116, 123, 132, 159, 180, 197
Advocacy work, 104-5
Affirmations, 29, 111
Anchor chart, 59, 70, 73, 95
Anxiety, 32, 76, 79, 81, 139
Art, 6, 22, 26, 111, 125, 136, 186-90, 196-97

Baking, 44, 51-52
Bar models, 66-67, 70-72
Base-10 number system, 117
Bayou Classic, 136
Bedtime Math Book series, 18
Bedtime routines, 31, 128
Biases, historical, 90
Black Girl MATHgic, 109
Black girls in STEM, 109
Black women, 109-10
Blocks, 2, 17, 87, 123, 132, 142, 183-84
Board games, 18, 22
Body-gons, 122-24
Born mathers, 15, 150
Brain breaks, 26, 79
Bumpy math journey, 35

Calming breath, 76-77
Calming glitter jars, 77
Candy, 98-99, 129
Card games, 18, 123
Cardinality, 37-38
CEO Math Trust, 54
Chant, 29, 162
Charts, 35, 74, 95, 101, 106, 130
Chatting, 165
Children's mathematics, 62
Choral counting, 39, 43

Choreography, 137
Claps, 26, 181, 185
Class communities, 94-95, 158, 175, 189
Cass helper, 165
Classmates, 5, 65, 73, 95-96, 117, 177-78, 181-82, 187, 190
Classrooms, 9, 24, 26, 35, 45, 61, 76, 79-81, 121, 123, 125, 148, 158-63, 195
Class rules, 172-73
Clink, 183
Clock, 123, 126-27, 129, 166
Clunk, 183
Clusters, 123
Cognitively Guided Instruction (CGI), 62, 75
Coins, 129-32
Collaboration, 5, 22, 57, 143, 147, 157-58
Collections, 41, 101, 130, 132, 188
Collective agency, 5, 11, 74, 174
College, 2, 11, 20, 24, 50, 107, 136, 164-65, 187-88
Colors, 12, 60, 98-99, 121
Combinations, 23, 118-19, 132
Comfort, 5, 116, 137, 145-48, 150-52, 163, 169, 196
Comfort B4 Confidence framework, 5, 146, 148
Common Core Standards, 63
Community of mathers, 5, 29, 44, 138, 147, 157
Community STEAM event, 40-41
Compassionate communication, 80
Competence, 147-48, 153-56, 158, 163
Composition, 40, 188
Comprehension skills, 17, 57
Computation, 117-18
Computation skills, 94, 133, 182
Conceptual understanding, 25, 34, 39, 75, 152, 154, 181, 190

Connections, 22-27, 51, 59, 65, 94, 103-4, 110-11, 118-20, 123, 132, 134-35, 164-65
Conservation of cardinality, 37
Content areas, 4, 31-32, 45, 73, 87-89, 91, 93, 95, 97, 99, 101, 103, 105-7, 109, 111, 113, 116, 153-54, 190, 197
Core academic skills, 15, 31, 33, 35, 37, 39, 41, 43, 45, 47, 49, 51, 53
Core math skills, 134
Core progressions, 25
Corner, 123, 129, 173
Counternarrative, 15, 21, 194, 199
Counting, 15, 34-41, 43, 99, 116, 118-20, 126, 130, 133, 135, 166
Counting coins, 130
Counting collections, 39-41, 43, 132
Counting objects, 16, 36
Counting principles, 36, 39
Counting skills, 39, 119
Counting strategies, 40, 63
Curiosity, 6, 16, 22, 32, 57, 60, 75, 91, 93, 115, 194, 197
Curriculum, 26, 93, 127, 129, 149, 151-52, 156, 161, 180
Cuthbertson, Ashley, 23-24

Dancers, 92, 134, 136-37
Data interpretation, 93, 101
Data representations, 95, 100
Data science, 101-2
Data unit, 94, 98
Difference, 13, 15, 76, 81, 94-95, 103, 129, 154, 167, 172, 176-77, 189, 196
Differentiation examples, 64, 70
Discriminatory practices, legacy of, 2, 4
Disney World, 136
Dot arrangements, 118-19
Dragon party, 64, 68
Dragons, 63-65, 67, 69-70, 72
Dragons Love Tacos, 63-64, 68-70
Dry lessons, 156

Early mathematics, 33
Eggo Waffles, 49
Elementary educators, 63, 87, 91, 111, 113, 146, 153-54, 157, 196, 198
Emotional reaction to math, 80

Engineering, 89-90
Environments, 16, 34-35, 81, 117, 123, 172
Equity, 181
Everyday problems, 19
Everyday tasks, 4, 12
Evidence, 37-39, 53
Exclusion, legacy of, 12, 199
Exploration in science to mathematics, 93

Failure in math, 165
Family math time, 31, 52
Family members, 17-19, 23, 75, 81, 95, 128
Fantastical stories, 14
Favorite activities, 133, 182
Feedback, 11, 26, 146, 177, 182
Filtering, 45-46
Finding solutions, 20, 59, 67, 154
Finger Movements, 118
Finger shaming, 117
Finger use in math, 119
Fish crackers, 13, 104
Formations, 136
Formulas, 24-25
Foundation, 14, 25, 35, 104, 111-12, 118-19, 121-22, 189
Fractions, 14, 19, 92, 97
Friendly fingers, 120
Function, 24-25, 95, 173
Fun stuff, 4, 115-17, 119, 121, 123, 125, 127, 129, 131, 133-35, 137-39

Game stations, 158
Geometry, 19, 123, 125
Girl math, 163
Girls in STEM, 109
Gonoodle.com, 79
Good habits of mathers, 45
Good learners, 4, 46, 61
Graphic organizers, 66, 72
Graphs, 95, 98-101, 106, 136
Grits, 48-50
Groups, 1-2, 39-40, 43-44, 60, 73, 116, 123-24, 132-33, 159, 161, 171-74
Growth mindset, 3, 5, 22, 29, 160, 169, 195

Habits of good learners, 4, 46, 61
Hard work, 20-21, 91, 116, 138, 145, 176, 197

Harris, Pam, 15
Home objects for students to count, 41
Homework, 18-19, 31, 41, 47, 74, 123, 125, 181
Hosting math nights, 161
Hour hand, 129

Inferences, 45-46
Information, 19, 45-46, 56, 63, 66, 92, 94, 100-102, 120, 162, 184-85
Injustice, 104-5
Instructional leaders, 3, 197
Instructional practices, 4, 146, 148
Instruments, 26, 79
Intention, 81, 190-91
Interconnectedness, 172
Intervention, 14, 29
Interview, 6, 23, 106, 109, 134, 164, 187

Jars, 77-78, 130-32, 162
Journals, 116, 179
Joy, 6, 17, 30, 83, 123, 189, 194, 197
Julia Robinson Mathematics Festival, 162
Justice, 99, 104-5

Labels, 72-73, 100, 129
Lacrosse, 157, 164, 166
Laib, Jenna, 14, 100-101
Language, 4, 13-14, 16, 29, 32, 53, 65, 129, 175-76, 194
Learners, 5, 16, 23, 44, 95, 109, 146, 151, 185, 199
Lessons, 5, 56, 60, 70, 81, 83, 99, 115, 117, 120, 126-27, 129-30, 152, 169, 180
Let's Share Stories, 55, 57, 59, 61, 63, 65, 67, 69, 71, 73, 75, 77, 79, 81, 83
Level playing field, 149, 151
Leverage, 32, 44, 46, 56, 83, 93, 138, 173, 196
Life lessons, 73, 81, 133
Listening, 66, 73, 79, 115, 181, 183-85
Listening error, 183-84
Literacy, 2, 26, 31, 34, 53, 56, 75, 81, 181-82
Literacy toolboxes, 61, 72
Little Flower Yoga, 81
Logic, 108, 120-21

Long-term academic success, 32
Lyrics, mathy, 53

Master of Business Administration (MBA), 108, 164
Math achievement, 32-33
Math and music, 27
Math anxiety, 5, 79-80, 91, 108, 111, 146, 195
Math block, 4, 9, 45, 57, 59, 85, 89, 138, 177, 180
Math brain, 1, 4, 108, 168, 181
Math by design, 174
Math circle time, 178, 181
Math clubs, 143
Math community, 5-6, 28, 160-61, 176, 190
Math community etiquette, 180
Math conversations, 196
Math crisis, 154
Math culture, 3, 146, 148, 153, 160
Math curiosity, 14
Math curriculum, 152, 169
Math discourse, 181, 190
Mathematical reasoning, 3, 152
Mathematizing, 59, 73, 86-87, 89, 91, 93, 95, 97, 99, 101, 103, 105, 107, 109, 111, 113
Mathematizing stories, 59, 83, 198
Mather, Queen, 110, 164
Mather Kids, 187
Mather Movement, 156, 163, 187, 194, 197-98
Mathers Math, 2, 4, 6-7, 10, 12, 14, 16, 18, 20, 22, 24, 26, 28, 32, 34, 36, 38, 40, 42, 44, 46, 48, 50, 52, 54, 56, 58, 60, 62, 64, 66, 68, 70, 72, 74, 76, 78, 80, 82, 84
Math experiences, 5, 32, 91, 116, 138, 146, 156, 197
Mathfirmation, 29, 53, 83, 113, 139, 170, 191
Math games, 19, 116, 162
Math gene myth, 21
Math geniuses, 21, 102
Mathing, 4, 6, 16-18, 21, 33-34, 45, 83-84, 87, 102-3, 134, 162, 182-84, 196
Math intervention, 112
Math intuition, 5, 13, 16, 150
Math joy, 3, 134, 138, 161, 182
Math Magical Fun Fest, 161
Math manipulatives, 151

Mathnote, 14, 21, 44, 61, 77, 100, 125, 157, 174
Math parties, 151
Math problems, 1, 20, 57, 66, 169, 197-98
Math progressions, 153
Math puzzles, 17-18, 21
Math riddles, 160
Math rotations, 123
Math routine, 31, 169, 181
Math standards, 63
Math stories, 5, 32, 57, 60-61, 63, 65-66, 70, 73-75, 83, 114, 161
Math tasks, 14, 18, 22, 33, 50, 52, 60, 80, 162
Math trauma, 9, 146
Micro-stories, 57
Mindful moments, 76, 127
Mindfulness, 76, 80-81, 144
Mindsets, 21, 29, 47, 53, 80, 83, 137-38, 168, 190-91, 194-95
Misconceptions, 12, 153, 194
Money jars, 131-33
Motor skills, 118
Moye, Michael, 80, 82
Music, 6, 14, 22-27

Next Generation Science Standards (NGSS), 93-94
Nickels, 129, 131
Number relationships, 18, 34-35, 39-40, 46, 67, 145
Number sense, 13, 15, 35, 86, 117, 153
Number sentence, 70, 72
Number sequence, 17, 36
Number symbols, 34, 119
Numeracy skills, 32
Numerosity, 13-14

Obsolete understandings, 45-46

Pace, 23, 116
Pancakes, 48-49, 95
Pathways, 111, 151, 194
Patterns, 4, 13-15, 22-25, 32, 39, 65, 102, 125, 137, 150, 185
Peart Crayton, Deborah, 54
Perplexors, 18, 120-21

Perseverance, 21, 91
Piñata, 60-61
Positive math experiences, 3, 5, 111, 173, 175, 177, 179, 181, 183, 185, 187, 189, 191
Positive math identities, 3, 5, 12, 15, 29, 32, 39, 91, 106, 146
Positive math identity development, 35, 111
Posters, 45, 99, 157, 185
Predictions, 9, 18, 60, 101
Problem of the day, 180
Problem-solving, 47, 61, 118, 162
Problem types, 63, 65-66, 73
Professional learning, 5, 33, 146, 148
Professional Learning Community (PLC), 65, 149
Projects, 11, 99-100, 143
Puzzles, 14, 22-23, 122, 162, 196-97

Quantities, 13, 34, 36, 39-40, 43-44, 58-59, 64-66, 70, 118, 120, 132

Readers Read, 4, 10, 12, 14, 16, 18, 20, 22, 24, 26, 28, 32, 34, 36, 38, 40, 42, 44, 46, 48, 50, 52, 54, 56, 58, 60, 62, 64, 66, 68, 70, 72, 74, 76, 78, 80, 82, 84
Reasoning, 62, 75-76, 108, 112, 120-21, 177, 179
  deductive, 120-21
  multiplicative, 35, 39
  proportional, 51, 63
Recipes, 20, 48-49, 51, 115
Rhodes, Brittany, 106, 110, 112
Routine, 60, 75, 95-97, 100-101, 113, 136, 149, 181

School community, 130, 156
School readiness, 32
School year, 94, 105, 155, 172, 175
Science, 31, 87, 90, 92-94, 99, 101-2, 108, 121
Science and math, 92
Sequence, 66, 72-73
Shapes, 12, 15, 61, 100-101, 122-25, 188
Silo, 26, 32, 89
Skittles, 98-99
Slow reveal graphs, 100-102

Social studies, 31, 87, 101-3, 121
Solutions, 5, 38-39, 57, 62, 64, 66-67, 70, 72-75, 94, 104-6, 176-77
Solving math problems, 52, 150
Solving story problems, 177
Solving word problems, 32
Splat, 38, 43-44
Squares, 183
Stability, 144-46
Standard order principle, 36
Standards, 16, 62, 88, 123, 134, 138
STEM, 21, 89-90, 92-93, 107-9, 111
STEM careers, 2, 4, 20, 90-91
Stereotypes, 2, 80, 146, 181
Steve Wyborney Estimation Clipboard activity, 162
Stories, 1-2, 10-11, 14, 33-34, 56-63, 65-67, 70, 72-76, 95-96, 109-10, 115-16, 125, 155, 162-63
Story problems, 58-60, 66, 74-75, 128, 133
Storytelling, 4, 57
Story time, 16, 31, 63, 73, 120
Strategizing, 59, 154
Students' voices, 75
Subtraction, 34, 63, 65-66, 119
Successor principle, 38-39
Summer programs, 40, 53, 81, 107, 186
Superpowers, sense-making, 5, 61, 106

Tacos, 63-65, 67-70, 72
Tang Math, 18
Tasks, 16, 22, 40-41, 46-47, 65, 86, 99, 116, 169, 180, 187
Telling time, 127
Thinking flexibly, 154
Think Space, 131
Triangles, 123, 125
Triple threats, 134

Vakharia, Vanessa, 27-28
Village agreements, 173-74, 176
Village mentality, 172, 176, 179, 190

Waffle Café, 48
West, Gladys, 100, 109
Whiteboards, 95, 128, 183-85
Word problems, 4, 32, 56-57, 63, 65, 76, 83
Workshops, 146, 154
Writers Write, 4, 10, 12, 14, 16, 18, 20, 22, 24, 26, 28, 32, 34, 36, 38, 40, 42, 44, 46, 48, 50, 52, 54, 56, 58, 60, 62, 64, 66, 68, 70, 72, 74, 76, 78, 80, 82, 84
Wyborney, Steve, 38, 150

Yoga, 76, 81
Yogis math, 77

CORWIN

To help every educator help every student

We believe   that every single student deserves a great education

We believe   that knowing our impact is both a privilege and a responsibility

We believe   that a fair, stable, and thriving society is built on education

Supporting Teachers, Empowering Students

Peter Liljedahl

Fourteen optimal practices for thinking that create an ideal setting for deep mathematics learning to occur.
Grades K–12

Peter Liljedahl, Maegan Giroux, Kyle Webb

Delve deeper into the implementation of the fourteen practices from Building Thinking Classrooms in Mathematics by focusing on the practice through the lens of tasks.
Grades K–5, 6–12

Rachel Lambert

Discover UDL for math, a way to design math classrooms that equips all students for meaningful and joyful math learning.
Grades K–8

Jo Boaler, Cathy Williams

Introduce data science to your students across disciplines with real-world stories and teacher testimonials to transform your classroom experience.
Grades K–8

Kimberly Rimbey, Katie Basham, Chryste Berda

Focus on making mathematics meaningful through multiple strategies and representations to help foster a love for mathematics in your students.
Grades K–6

Sherri Martinie, Jessica Lane, Janet Stramel, Jolene Goodheart Peterson, Julie Thiele

Build critical intrapersonal and interpersonal skills *through* mathematics to help all students grow the life-skills they'll carry forever.
Grades K–12

To order your copies, visit corwin.com/math

CORWIN Mathematics

Our research-based and high-quality content is written by trusted experts and provides clear pathways to helping all students gain access to rigorous mathematics learning; to learn to truly think, reason, collaborate, and fluently discuss mathematics; to form positive and strengths-based mathematical identities; and to see and use mathematics as a tool to effect change in their lives and communities.

Sue Chapman, Holly Burwell, Mary Mitchell

Focus on developing eight essential habits that support mathematical competence and confidence in students through a personalized, practice-based professional learning experience.
Grades K–5

Pamela Weber Harris

Guide students through domains of mathematical reasoning, from counting and adding strategies to more complex proportional and functional reasoning—without resorting to algorithms.
Grades K–12

John J. SanGiovanni, Dennis McDonald

Enhance your students' understanding and engagement in geometry, measurement, and data while also fostering a deeper connection between math and the real world.
Grades K–5

Vanessa "The Math Guru" Vakharia

Equip students to develop the skills they need to truly believe anything is possible, even a better relationship with math!
Grades K–12

Liesl McConchie

Reexamine what it means to have a positive math identity—and learn to use brain-based tools in a humorous and friendly way to build on a positive math identity for your students.
Grades K–12

Irina Lyublinskaya, Xiaoxue Du

Integrate AI literacy into K–12 classrooms, blending theory, practical lesson plans, and ethical considerations to empower students as critical thinkers.
Grades K–12

To order your copies, visit corwin.com/math

CORWIN Mathematics